To see a World in a grain of sand,
And a Heaven in a wild flower,
Hold Infinity in the palm of your hand,
And Eternity in an hour.

William Blake

THE GREAT SEASONS

presented in words by

DAVID BELLAMY

and in pictures by

SHEILA MACKIE

Hodder & Stoughton

LONDON   SYDNEY   AUCKLAND   TORONTO

In association with Oriel Press

British Library Cataloguing in Publication Data

Bellamy, David
   The great seasons.
   1. Evolution
   2. Natural history – England – Teesdale
   I. Title   II. Mackie, Shelia
   575   QH366.2

ISBN 0 340 25720 2

Designed and produced in Great Britain by Oriel Press Ltd. at Stocksfield Studio, Stocksfield, Northumberland, NE43 7NA

Typesetting by Knight and Forster Ltd. Leeds

Colour separations by Spectrum Photolitho, Bradford

Printed and bound in Great Britain by
William Clowes (Beccles) Limited, Beccles and London

Published by Hodder and Stoughton Ltd., Mill Road,
Dunton Green, Sevenoaks, Kent

Editorial office: 47 Bedford Square, London WC1B 3DP

To
Margaret E. Bradshaw

Acknowledgements

Our thanks to:- Rev. Gordon Graham, Wendy Tickle, David Shimwell, Clive Marshall, Peter Bridgewater, Peter Holland, John Peters, Peter Davis, Sandra Nye, Trevor Elkington, Joe Harvey, all students and colleagues who have helped me in my understanding of the Upper Dale: Major G.L. Lister and Don Wilcox for help with matters archaeological: Dennis Rowlands, a good friend who has a very special part of the Upper Dale in his care: the Misses A. and D. Redfearn for allowing me to quote freely from the manuscript of *The Redfearn Saga*: Dorothy Dover, T.C. Dunn, June Hodges, Roy Tyerman and John Young for help with research and materials.

We wish to thank Bruce Allsopp for his vision and patience in bringing us together and this work to fruition.

D.B. & S.M.

Foreword
by Her Majesty Queen Elizabeth
The Queen Mother

I am delighted that a new book, THE GREAT SEASONS, has been written about the dales of the North East of England, describing that wonderfully interesting and varied countryside and the people who have contributed so richly to its history. It is a part of England very close to my heart and I have the happiest memories of the time I spent there in my youth.

This book with its beautiful illustrations will, I hope, be a source of very real pleasure and enjoyment to readers of every age and background.

Contents

A Beginning

E know a place where, with care and knowledge, it is possible to reach down through the whole history of landscape back to the time when it was a white desert of ice and driven snow. This place lies deep in a fold of the great hills – fells we call them – which form the upper catchment of a river in Northern England. It is a place of environmental magic, where you can make use of all your senses to review the course of the world's history.

To *feel* history is a very special thing and there the cool peat, black-brown, soft to touch, gives way under press of eager fingers. Down they go, down through the leaves of the past, each layer older than the last, until they reach the cold resilience of clay that shapes and holds the feet of all the highest fells: clay that was laid down by glaciers whose solid substance has long ago joined the eternal cycle that makes clouds from oceans, fills rivers and lakes and waters the dry earth.

An incessant melody of noise is there, the rill of water tumbling down the beck, as mountain streams are called in these parts, suffusing through warm limestone to emerge crystal clear, the perfect habitat for a close relation of the red seaweeds which clothe the distant shores of the North Sea to which the water always gravitates. This, together with the hiss of wind through Heather, gives background to these historical events.

The sense of smell awakens as the peat is stirred, bringing up the odours of semi-decay. Sulphuretted hydrogen (bad eggs to most), spells out the lack of oxygen and gives reason to the perfection of the peaty record. This is no Chinese egg, a mere thousand years of age but a profile rich in records of the past, twelve thousand years of history to be savoured.

1

To feel, to hear, to smell; yes, if you have the nerve, here you can even get the *taste* of events long past. The upper peat is bland, soft, smooth as paste, but then below are layers sharp to the tongue and packed full of grit washed down when Man or climate, or perhaps a combination of the two, removed the forest and released rich soil from bondage of its roots.

The fifth of our sharp senses is that of sight, and here the great book of the fells will only disclose its deepest secrets to those students of science whose dedication allows them to probe with powerful instruments into the myriad miniature records of grains and spores and cast off carapaces.

The living tapestry which clothes the mighty fells records the high-lights of the past events which shaped this living landscape. The players in this pastiche of history are the plants that experts call the Rarities of the Upper Dale. Each one a gift of the Four Great Seasons, past climates that left the dale enriched with beauty, each one is now held fast in the unique environmental crown.

False Sedge, dark present from a winter realm where life is harsh amongst the crystals of cold water. Spring Gentian, blue as any heaven on an alpine day, a gift from the kingdoms of great mountains. The Mealy Primrose, a pink powderpuff, this tells of life in alpine hinterlands where spring is soon eclipsed by summer sun. The sweet presence of Bitter Milkwort breathes the full cadence of warmer southern realms, while deep purple arils drip their autumn scent from the topiary of Juniper.

All these, and many more, are there for all to see, to smell, to taste, to hear, to sense the story of the Four Great Seasons which have shaped this and the majority of the landscapes on which the world of modern man has come to depend. This is the story of one very special place, but in its telling there is deep meaning for all who want to have the sense of history and, in a moment of understanding, find the roots of their past, the reason of their present and sense for their future.

The place has a name but we will simply call it Upper Dale. The reason for this decision will become obvious as the story unfolds.

Down through the peat.
  Fossilised trees, Auroch horns and early human artefacts
  Common Frog, Fox, Harebell, Cross-leaved Heath, Bog Asphodel

PLATE 1

Tilting at Sunbeams

PRING, SUMMER, AUTUMN, WINTER, four seasons which shape each living year, at least in all those regions of the earth that lie beyond the compass of the tropics.

Between Capricorn and Cancer each day approximates to twelve hours' light, each night to twelve hours' dark, a cycle which remains invariant throughout each year for it is determined by the spin of the earth upon its polar axis. The poles stay cold and in essence lifeless while the rest of the globe is turned before the fiery sun like some gigantic ball of meat set on a cosmic spit. Each segment of the earth and its atmospheric envelope is thus warmed throughout the day and cools throughout the night and it is this rare combination of rate of spin and spatial distance from the sun that makes life as we know it possible.

If the earth did not spin, each day and night would be some six months long, determined by the passage of the earth around the sun, and during those long nights and days the temperature would range far beyond the limitations of the living state and the whole planet would be as lifeless as its poles.

If the earth did not rotate around the sun there would be no sidereal year marking the passage of three hundred and sixty five and one quarter days. Likewise if the earth were not tilted by twenty three and one half degrees upon its polar axis there would be no seasons to put the main pulse into the life of its hemispheres.

The earth in its measured quixotic journey through space thus tilts at sunbeams and it is this tilt that brings about the differential heating of the hemispheres. From March 21st each year for six whole months the tilt ensures that

3

the north is more inclined towards the sun and summer warmth creeps up even to the pole. For the next six months from September 23rd the same is true of the south and there summer attains its warm solstice, its longest day, on December 21st. From that point on the southern days shorten towards the equinox when on March 21st the sun stands once more directly above the equator and the daily round of heat and light across the globe is the same as it would be throughout the year if the earth were not tilted on its axis of rotation.

If the earth were of a different mass the atmospheric blanket would differ both in thickness and in composition: it would thus either shut out or let in more or less of the radiation of the sun, both in respect of heat and light, and life as we know it would never have developed. The magic blanket does not only provide us with the oxygen we breathe, but it protects us from the worst effects of both cosmic and ultra violet radiation, acts as a greenhouse holding in the warming rays of the sun and gradually re-radiating them to outer space. The overall density of the blanket is fixed by the mass and hence the gravity of the planet, but its composition is at present held in delicate balance by the processes of the living world. If that mass or that balance changed? But these are not ifs and buts, they are facts which concern this one small corner of the heavens and as such they are the facts of life. For it was against these fixed functions of mass, orbit, spin and tilt that life evolved, and it is these same facts and functions which mark the passage of time through which evolution has wrought its marvels including you and me, the world of conscious, thinking men and women.

Thirty days hath September, April, June, and November.
All the rest have thirty one,
save February alone which has but twenty eight days clear,
and twenty nine in each leap year.

Thus it is that the true cycle of the sidereal year is adjusted to fit Man's calendar of events, a matter not of scientific fact but of pure convenience to accommodate the seasons. For in the matter of the seasons there is no room for convenience; Man like all other members of the natural world must adjust his way

Ice floes in the waters of the Dale

Young Polar Bear   *Starry Saxifrage*   *Barnacle Goose*

PLATE 2

of life to fit in with the temperature of the tilted world. Why?

Well, there is a law of nature which dictates that the rate of all chemical reactions is dependent on the temperature at which they take place. As a rule of reaction it may be stated categorically that 'a rise of ten degrees on the scale of Celsius (10°C) will bring about a doubling of the rate at which a reaction will proceed'.

Now to whichever line of origin, apes or angels, evolution or creation, you lend your conscious support, there is, as far as I am aware, no body of thinking which denies the fact that the living state is in essence a complex of chemical reactions. The most ardent supporter of special creation knows he or she requires chemical food which is put to good use in his or her chemistry of living. In such guise life must obey the rule of reaction, quickening and slowing with every rise and fall of the thermometer. Thus it is that each seasonal shift in temperature is mirrored by a shift in the pace of living.

To date the world's highest temperature on record is 58.0°C and it was reached on September 13th, 1922 at Al Aziziyah in Libya and the coldest, −88.3°C in Vostok in Antarctica on August 24th, 1960. These are the current official world extremes of temperature. However, in the knowledge that official meteorological stations are few and far between and that their records are taken from instruments set in the shade of a well ventilated box resplendent in an oft renewed coat of white paint, the absolute range must be more extreme than 146.3°.

Anyone who has leant against a motor car which has been left standing in the sun will appreciate the difference between sun and shade temperatures, and the darker the paint on the car the more rapid will be their appreciation. Black bodies absorb heat more efficiently than white ones, the simple reason being that the whiter the body the more it reflects the sun's energy, and that includes the full spectrum from the ultra violet down to the infra red (heat) rays.

When talking about radiation, physicists usually refer to 'the perfect black body'; now whether that is a Rolls Royce or a Cadillac, I don't know, but what I do know is that temperatures which would be lethal to most forms of life are of common occurrence across the face of the globe.

I must, however, add that one of the major effects of a rise in temperature is to bring about an increase in the rate at which water evaporates. As water evaporates it changes from a liquid to a gas and this change of state uses up a lot of energy (the latent heat of evaporation), cooling the body, be it black or white. It is not just the prerogative of *Homo sapiens* to keep himself cool by perspiring or herself cool by glowing, but as all life consists of at least 70% of water, all life has an inbuilt cooling system as it were on tap, and can thus survive, be it for very short periods, when exposed to theoretically lethal temperatures.

It is interesting to remind ourselves at this juncture that animals can, in the main, move into shade when things begin to get too hot whereas plants have to stay put and 'sweat' it out. The loss of water from land plants, which can be of great importance in keeping leaf surfaces down to reaction temperature, is called transpiration and much of the evolution of the plant kingdom has revolved around the problem of conserving water and keeping cool, both at the same time.

At the lower end of the temperature spectrum the limits are a little easier to define because life comes to a pretty solid halt at around 0°C, the temperature at which pure liquid water becomes pure solid ice. Life and ice, or, come to that, ice and any other chemical reaction don't mix. Water is the universal solvent, the medium in which the vast majority of all chemical reactions take place. It is also the universal pep pill (catalyst) which speeds them on their way.

In order to react chemicals have to be brought together and, in the absence of a chemist with a stirring rod, mixing is brought about through the agency of diffusion. Diffusion through liquid water is a slow process, being the movement of chemicals from an area in which they are more concentrated to one in which they are more dilute. As the chemicals move in one direction water moves in the other, the whole process smoothing out the gradients of concentration. Diffusion through solid water is a much slower process for the water molecules themselves are held fast, part of the lattice structure of the ice.

Ice crystals have a second very drastic effect on the living chemistry which is

ARCTIC FOX

PLATE 3

itself ordered in minute structures called organelles. It is on the integrity of these organelles that much of the living chemistry depends – ordered life chemicals set in a labile matrix of water which itself can freeze, slowing the process of diffusion and the crystals themselves disrupting the structure of the organelles, causing damage which must be repaired before the living chemistry can once again get into top gear.

The third effect of solid water is perhaps too obvious to dwell upon at any length and yet it is the most far reaching in its effects. Solid water forms a blanket of white snow which under the pressure of fresh falls turns into ice. The effects of this solid layer are manifold.

First it acts as a thermal blanket, insulating and thus protecting anything beneath from the colder temperature which may range above. Organisms able to survive the rigours of zero plus a little may thus hibernate in safety beneath a layer of snow or ice. It is, however, unfortunate that the reflective nature of the thermal blanket ensures that much of the light energy is re-radiated back into space. This means that the periods of hibernation are strictly limited by the amount of energy the organism is able to put into store.

A third problem is that a layer of solid water effectively shuts the living system off from the supply of minerals in the soil beneath. Any plant which bridges the gap, with its leaves above the snow and its roots fast in the frozen soil beneath, will tend to lose water at the top and be unable to take it in, in solid form, through its roots and will, in effect, dry to death.

The upper and lower temperature limits of life on earth are thus set by the interplay of temperature and the states of water and, between the extremes, the seasons hold their sway.

Living in twentieth century affluence, protected from the vagaries of climate in homes that may be heated or cooled at our convenience, it is so easy to forget the effects of the tilted earth. Caves, animal skins, weaving, the use of fire and now the whole panoply of modern technology have made us masters of the world's environments, putting us, it would seem, beyond the reach of the seasons.

However this is not so. Just think how each winter brings to light the problems of transport, each autumn demonstrates the rising cost of fuel for heating and, last but not least, consider the fact that much of Man's world looks towards the rich wheat belts and rangelands of the temperate zones for the bulk of the food we eat. To date extensive agriculture and husbandry in the tropics has met with less success. The reasons are manifold but, in part at least, they centre on the lack of seasonality determined by temperature.

Although winter is usually looked upon with trepidation it must be remembered that it is not just a time of death and destruction but one of rejuvenation and cleansing also. Plants die back and as they do the parasites and diseases they have harboured during the growing season themselves have to hibernate or die. In the absence of a thick protective cover of living vegetation frosts can stir deep into the soil, helping to break down the particles of rock and the dead organic material, releasing new and recycling old stores of nutrients ready for the new growth of spring. The winter initiates the cycle of the elements that are such an important part of every living system, cycles of matter within the cycle of the seasons. The seasons are thus pacemakers on a massive scale. Life sleeps, wakes, lives and slows once more, back to rest in winter beneath a soft white blanket of snow.

In the tropics and sub-tropic climes respite from the daily round of growth comes only with periods of drought when lack of water brings the living process to a halt until sufficient rain begins to fall once more. The wet and dry seasons are much more an on/off switch, an all or nothing, than a pacemaker varying the pulse of living. In those regions where there are neither seasons nor wet/dry periodicity each plant and animal may follow its own inbuilt rhythm of life, each out of phase with the other. The result is that a constant drain is put on the resources of the soil, a drain which has, over thousands of unchanging years, sapped the mineral reserves within the sphere of rooting; so much so that today there are no reserves waiting to be used if Man removes the forest and tills the soil to his own ends. So it is that agricultural Man has found and always will find great limitations within the tropics, except in specially favoured places. In such understanding lies not only

The tongue of the Glacier

PLATE 4

the secret of the temperate climes, but the story of the Four Great Seasons.

The annual march of the seasons is however set against a background of much more ponderous change of climate. Although the tropics have enjoyed their constant warmth over an immense span of time the regions nearer the poles have known much colder times which have resulted in the development of great ice sheets. It is strange to think that in the two million years during which Man has stood erect upon the earth the higher altitudes and latitudes of Eurasia and North America have been engulfed by ice on no less than four occasions. Each period of glaciation may therefore be regarded as a Great Winter and though each brought death and destruction on a massive scale, they cleansed and prepared the landscapes ready for the Great Springs that followed each one, initiating new cycles of living.

The last of these Great Winters reached its zenith around 18,000 years ago and then began its slow decline as the green fingers of the last Great Spring spread across the new landscapes. It was into the richness of the new season's growth that Man walked erect into the realisation of his potential. As the last Great Spring began the earth supported some four million human beings on the affluence of the land between the sheets of melting ice. Today it supports four billion of the same species, a thousand fold increase, a story of unprecedented success, much of which depends upon the affluence of the new landscapes that were prepared by the ice sheets of the last Great Winter.

The Making of Landscapes

ATER VAPOUR is held in the form of clouds by the gravitational pull of the earth. Once the water vapour has changed its state, condensing to form drops of liquid, these obey the law of Newton and fall as rain. Each droplet once formed has its own store of potential energy which is a function of its weight and its distance from the earth's own centre of gravity. This potential, or at least part of it, will be dissipated on its way down; no wonder then that hailstones hurt, especially when one can weigh a record 750 gms!

One gram of water falling on the highest point in the Upper Dale, which is 893 metres above the level of the sea, has the equivalent of a little over 89 joules of potential energy which could be given up as it continues on the way of gravity down towards the cold North Sea.

Since the records began each hectare of this fell has received in an average year around 23,000 tonnes of rain, its gravitational potential 56,200 kilowatt hours which is equal to the energy in 300 barrels of crude oil. Averaged out for the whole Upper Dale catchment the figures become impressive; 427 million tonnes of water, 1,045 million kilowatt hours of potential energy equivalent to 5.5 million barrels of crude oil.

Bucket, or is it barrel, mathematics of this type must be grossly inaccurate but it's fun and it does show, in terms which are immediately understood by our switch-on, gas guzzling twentieth century society, the potential power of the water that daily flows through the catchment system on which you live. (I have worked it out for part of my local catchment; I suggest you work it out for yours.) All this energy must be dissipated as the water makes its way down to the not so mighty

10

The retreating Glacier. Sabine Gulls

PLATE 5

North Sea which is, at the time of writing this chapter, supplying Man's energy-hungry economy with 1.6 million barrels of crude oil each day.

Crude as these figures may be it is vital to remember that the potential which fills both these energy sources is the same: the energy of the sun stored by vegetation long since dead, and the energy of the sun stored in much more recent times as the convection currents in the atmosphere lift the water vapour up to form clouds above the catchments. The one part of the deposit account of world energy is a non-renewable resource on which we have come so to depend. The other is a tiny part of the world's current account of energy to which we must turn more and more in the not far distant future.

Long before Man came into the Upper Dale scene (or come to that on any other landscape) those energy statistics were there; one source hidden below the waves waiting to be put to use; the other being dissipated in re-evaporation of water, traction of particles, friction, scouring and dissolution of the catchment. This is why water draining through a river system eats into the terrain, its erosive effect being concentrated, as run-off and seepage join to form rills, streamlets, streams, tributaries, and eventually flow into the main body of the river itself. The end product was a network of V shaped valleys cutting back into the hills, a water worked landscape. So it remained until the coming of the ice of the last Great Winter.

Freezing stops this rapid cycle of erosion dead in its liquid tracks, covering the terrain first with a blanket of white reflective snow which under the pressure of fresh falls turns to ice. In a normal year this pent up energy will be released on its erosive way at the coming of spring.

In the Upper Dale the ice sheets of the last glaciation built up to their maximum over a period of about 3,000 years. Now let us suppose, in ignorance of the exact data, that over that time precipitation in the Upper Dale equalled the average now recorded on the fell side, some 150 cms per year, and that it all stayed put exactly where it fell. The thickness of the ice sheet would have approximated to 4,500 metres, half the height of Mount Everest. There is good reason to believe that this figure is wildly inaccurate and that the ice sheet was nowhere near that thick.

The reason is not that the annual precipitation was greatly reduced during the period of glacial advance but that the onset of the glaciation did not bring the normal march of the seasons to a halt. The 'permanent' covering of ice smothered all life beneath but above its icy presence the cyclic seasons went their annual way.

At the onset of each winter new falls of snow crystals are added to the top of the ice, entrapping air as they accumulate. If nothing further happened the snow blanket would build up very rapidly, for 12cms of freshly fallen snow equals a mere 1cm of rain. Remember what happens when you make a snowball: you squeeze out the air and turn the powderpuff of snow crystals into a ball of ice. Likewise the weight of fresh falls compresses the snow-air layer into firn or névé which, largely under its own weight, compacts further to form ice with isolated bubbles of air entrapped in what is otherwise an impermeable matrix of solid water. It is thus that glacier ice is formed, a process which is speeded up by surface melt during the spring and summer and by the addition of more ice which is formed when water vapour in the form of mists and clouds freezes on contact with the ice surface. However, it is also during this process of summer melt that some of the water is lost to gravity and some to evaporation.

All that had happened was a shift in the equilibrium of the seasons, a shift which favoured the formation of glacial ice over a large section of the globe. Such a shift need not have been very great and evidence indicates that the average annual temperature in Britain over the period of glaciation was only some 5°C less than that which we tolerate today.

The theories as to exactly why this shift took place and why other similar shifts have taken place in geologically recent times are too numerous to detail. However, let us wander into one flight of fancy and suppose that the earth has, on a number of occasions, tilted further upon its axis of rotation, rocking back to take up its more normal position in relation to the sunbeams. Certainly such an increase in tilt would mean that the winters were colder as the warming rays of the sun

Receding Ice Sheet with Greylag Geese

PLATE 6

would strike the upper latitudes at a lower angle. The converse would also be true and the summers would be warmer, melting the ice and nullifying the effect. However, the latter is not necessarily the case, for the extra heat of summer could evaporate more water from the surface of the oceans. The resultant banks of cloud would protect the ice sheets from the direct rays of the sun, for water vapour absorbs infra red heat rays with great efficiency. Likewise the oncoming of each harsher winter would flush away the clouds, for ice crystals act as nuclei for precipitation and snow would fall upon the firn beneath.

If the thought of an unstable earth drives you towards disbelief I would remind you of the following facts. In many areas of the more northern climes the land surface is rising beneath your very feet. The reason is that now the immense weight of the ice sheets has gone, the earth's crust is, as it were, heaving an elastic sigh of relief. Add to this the fact that we now have good evidence that the continents themselves are on the move, floating across the face of the planet. Each continental plate is pushed by the formation of new crust spewing forth from volcanoes deep in the mid-ocean trenches. In fact, to push the question of what brought about the change in the balance of the seasons back one stage further, it could be this movement of the continents which has caused the earth to rock on its axis, like some giant flywheel somewhat out of balance.

These explanations may be no more than cosmic codswallop but it doesn't really matter for it is the effect of the Great Winter, not its cause, that has shaped the world as we know it. The facts are there etched across the face of the temperate landscapes of the world for all people to see, especially if they have the desire to take note and understand.

On a number of occasions over the last two million years ice sheets have developed from the highest latitudes and altitudes of earth, wiping out all life in their compass, pressing down on the earth's crust and reshaping it by their slowly moving presence.

Modern glaciers appear to be of two distinctive types, the ice cap or ice sheet which extends as a continuous layer obliterating the waterworn landscapes beneath, and the valley glacier which, as its name implies, part fills a valley from the bottom upwards and thus is contained in a rock-walled trough. Yet it is easy to see that one is simply an extension of the other, both in space and time, and that any ice sheet can both start and end its life as a series of valley glaciers.

As the glacier in the Upper Dale built up, not only did it begin to override the margins of the valley but it also began to move downslope under the force of gravity. The first small corrie glaciers of the Upper Dale thus formed nuclei from which the ice sheet gradually built up, moving both downslope and down latitude, joined and overridden by glaciers formed outside that valley.

But how does a mass of ice move? As more firn is added at the top the overall pressure increases and the ice of the valley glacier becomes plastically deformed and begins to move by a process called creep. It is rather like toothpaste being squeezed out of a tube, though in this case the tube is the size of the valley and the ice comes out not as a worm-like ribbon but in the form of spectacular spreading lobes. The process of creep is very slow but nevertheless it is of great importance, especially during the expansion of the valley glaciers which will eventually coalesce to form an ice sheet.

A second mechanism of movement is when the ice comes under such strain that it shears along lines of weakness rather like rocks shear to produce faults. The faults so produced are called crevasses and the rate of shear movement may be both rapid and spectacular.

The third and often the commonest mechanism of movement is called basal slip and this appears especially active in glaciers in warmer conditions like those that would have prevailed towards the end of the period of glaciation. Basal slip probably only takes place when there is a film of liquid water present between the base of the glacier and the rock surface across which it is moving. The presence of liquid water within the ice appears to be due to the pressure set up by the enormous weight of ice above which alters the freezing point of water. Only a tiny amount is needed to lubricate the base of the glacier, allowing movement down-slope. So it was that the potential of all that pent up solid water pressed down on

Graptolites

Geologist's hammer

PLATE 7

the old waterworn landscapes, gouging out a new valley fit to hold a river of ice, and a glacier, however large, is just that.

The fact that such things happened can be seen in the U shape of glacial valleys still in the making and the behaviour of the glaciers they contain. Likewise it has been found that the scratchings on recently exposed rocks mark the direction in which the ice has travelled, each one chiselled out by lumps of rock held fast in the underside of the ice. Thus it was that the solid water of the last Great Winter put a new U shaped face on the valleys it covered, reshaping the work of erosion that liquid water had done through warmer aeons past.

As always, the rate of removal of rock is a function of the frictional energy applied, be it by ice or water, and the character of the rock itself, especially as regards its softness or hardness.

The bulk of the rocks from which the landscape of the Upper Dale was formed, was shaped by the action of water both solid and liquid, during the carboniferous period which lasted for 165 million years and came to a close some 280 million years ago.

At only one place in the whole of the Upper Dale do older rocks dating from 500 million years ago come to the surface. They outcrop close beside the river to form what the geologists call the Upper Dale inlier. It is a very special place, a spot not where time has stood still but where the record of 500 million years of geological time has been swept away by the master sculptor erosion, uncovering secrets of an earlier age.

I like to stand there close by a tiny quarry called Pencil Mill where soft green slate was hard won to make the slate pencils for a bygone age of scholarship. There I can travel, at least in my mind's eye, back across the years to the time when the slabs were laid down, a time when the only forms of life were in the sea and all the land was a desert of bare rock.

Just as the children of Upper Dale had used slate pencils from the quarry to mark their progress in learning so too had the inhabitants of that age much further off left their own record in the same green slates, a record of the progress of evolution which would be read by learned men and women of the distant future. The frail imprints of Graptolites are there to see, members of a group of animals long since extinct which were related in some respects to our modern jellyfish and corals. In the right hands these allow not only a glimpse of the state of play of evolution of the animal kingdom but allow this small pocket of history to be placed in its correct sequence of time, the rocks themselves spelling out the conditions under which they were formed.

The same is true of the great bulk of the carboniferous deposits which form the ramparts of all the rest of the Upper Dale. Although they are in the main hidden by more recent deposits, they are there, a firm foundation of history, a great stepped pyramid which was almost 65 million years in the making.

The story which these rocks tell is of a sea set under tropic skies somewhere south of the equator: a sea teeming with life surrounded by land on which green plants akin to our modern ferns grew and amongst which great amphibians tested out their landlegs and winged insects flew. At that time evolution was going on apace and the record of its progress was laid down in the thick deposits on the floor of the sea, sediments which in time were compressed to form a distinct series of limestone rock strata. It was a shallow sea, an ocean in the making, and at times the rate at which new sediments were being laid down was faster than that at which the floor of the basin was sinking. So it was that sections of the sea became semi-dry land and these were immediately colonised by the amphibians of the plant world, the clubmosses, horsetails and ferns. Though many were much larger than the ones we know today these giants of the past had to live with their woody stigmarian feet, as it were, in water for, like their modern day counterparts, they could not complete their life cycle in the dry.

The term carboniferous means coal bearing and to prove it the remains of these tropical swamps are there, thin bands of coal sandwiched between the lime rich rocks. The dark colour of the coal is due to the abundance of carbon which made up the trunks, branches and leaves of the coal forming plants. The sediments which formed the limestones themselves, though equally dependent on

Canada Goose and Snow Buntings.

PLATE 8

the pulse of life in the seas, contain less organic carbon and hence the rock is lighter in colour. Carbon is nevertheless present, locked up as carbonates of calcium and magnesium which, being insoluble in the water of the sea, joined the constant rain of material down from the sunlit productive surface waters into the darkness beneath.

The purest of all the limestones lies almost at the base of the pyramid. It is light grey in colour and bears the name of Melmerby, a village many miles outside the Dale, where it was first described. All the other limestone strata in the Upper Dale are less purist in their mode of formation and in their substance.

A shallow sea beset with land from which rivers pour forth silt, mud, sand and pebbles, must bear witness to these facts both within its life and in what it leaves behind. Thus, apart from the thin coal seams, the limestone steps of the pyramid are interspersed with impure limestones, shales and sand, each of varying colour, texture and resistance to the passage of time and water in whatever form. The softest in the series are the shales and the hardest the pure limestone and the sandstones. The latter were laid down towards the close of the carboniferous times and form the caps and scarps of the highest fells.

Thus it was that the foundations of the Upper Dale were laid down in the same element — water — that would much later act as master sculptor and create the form of the landscape as we know it.

Today most of this great pyramid of carboniferous time lies hidden beneath deposits of a much more recent nature, all dating back no further than 20,000 years, to the time when the local ice sheets of the last Great Winter were beginning to form. The story of the rocks has been unearthed by mining in the Dale and pieced together by the many learned people who have tramped the fells, split rocks with hammers and sat for many patient hours studying the minutiae of their detail.

It is they who have dated the sequence of events and shown that this particular pyramid of time floats in limbo between two great voids. One hundred million years of time before the carboniferous period began and two hundred and fifty million years since its close are missing, wiped away by erosion and by events of cataclysmic magnitude that have occurred as this one piece of continental crust has made its way across the face of the globe from the tropics to its present temperate location.

Continents have drifted, seas have come and gone, mountains have been formed and eroded away, but these rocks have remained in their ordered sequence and it is against their background that our story will unfold.

There is, however, one more feature of these rocky ramparts which form not only a cornerstone to the foundations of the Dale but also to our story; its name is the Great Whinsill. At some time in the past the movements of this particular section of the earth's crust cracked the rocks, opening up fissures which reached down to the molten magma upon which the crust floats. New igneous rock (molten lava) began to pour up towards the surface where it was injected into the cracks between the bedding planes of the limestone and shales, to form, as it cooled, sills of rock, fluted columns of quartz dolerite, reminiscent but not so regular as those of Fingal's Cave and the Giant's Causeway. This is the hardest rock of all those found in the Upper Dale and as such has stood the test of time, playing an important part in the special magic of this landscape, a magic which makes it stand above all others as the best in which to recount the story of the Four Great Seasons.

Wherever the molten lava touched the bands of coal the fossil fuel was consumed and its acrid smoke must have joined with the vapours of volcanic action seeping to the surface. Likewise where the lava touched the limestones and the shales it baked them, changing (metamorphosing) them both in form and makeup. Sandstones became white quartzite, metals melted and ran together producing rich veins and lodes, shales turned to whetstones, the impurities in the limestone were either burned out or were metamorphosed into new minerals, some, like garnet and iriocrase, of great potential beauty. The pure limestones were themselves baked into marbles and the purest of all into a coarse crystalline mass which now bears the name *sugar limestone*, a deposit which was to play an important role in Upper Dale. Out of the strength of volcanic fire came forth a

The Foundations of the Dale.

PLATE 9

special sweetness that would, some 295 million years later, flower to great importance.

The developing glaciers thus began to form, bounded by this multidecked rock sandwich, some of which had been baked to perfection from within. The scene was set for the final transformation.

The glacier which filled and then overflowed the margins of the Upper Dale can be looked upon as a gigantic file, shaped to fit the terrain which moulded its formation. As it moved forward and down under the force of gravity it began its work of shaping the then hidden landscape. The major mechanisms of erosion are abrasion, which polishes and scratches the rock surface, and plucking, which though much more complex in its action removes larger lumps, although in the majority of cases these will have been loosened first by frost shattering, a mechanism which makes use of both the liquid and solid states of water. Meltwater infiltrating into cracks re-freezes and it is the resultant increase in volume which flakes pieces of the rock away. The adhesive strength of ice on even bare rock will always be less than the shear strength of the rock itself and thus it is very unlikely that ice plucking of intact rock ever occurs. Although almost impossible to measure, the rate of lowering, that is erosion over a large surface, has been calculated to be at maximum of 2.8 mm per year, although 1 mm would perhaps be a more normal average figure. Nevertheless, continuous loss at that rate over a period of 5,000 years could bring about a lowering of landsurface of some 50 metres, a rate almost twenty times as fast as that produced by the action of liquid water in a normal river system. Held by the original forms of the rock the glacier grinds out the basic shapes of the new landscapes.

The best evidence for massive erosion is in the form of glacial overflow channels which cut north east out of the main valley of the river. Meltwater ponded up beneath the glacier must have helped the ice to breach the valley walls, spilling the water into the adjacent river systems. They are easy to recognise; rugged valleys that must have been carved out by an enormous force of water now harbour in their V shaped depths the merest trickle of a stream. They have no real catchment of their own, being vestiges of an earlier age. They are hanging valleys which bridge the catchments of the two ajacent rivers.

Glacial erosion within the Upper Dale is perhaps best described as cosmetic surgery rather than a complete restructure job. This points to the fact that the ice sheet was never very thick above the fells and/or that its overall movements were limited.

The products of ice erosion, however fast or slow, consist of rock fragments ranging from large lumps to flour, the latter consisting of particles so fine that they can only be seen with the naked eye when suspended in water which they turn to 'milk'. These erosion products may be carried along frozen to the underside of the ice, new teeth in the fitted file, they may be bulldozed along by the ice or may be carried along more rapidly by the meltwaters emanating from beneath the glacial mass. As all these rock smashing forces are at work throughout the life of the glacier right from its birth to its demise, the ice mass is itself far from pure, for it carries many of the products along in the form of what are called erratics, although this term is usually reserved for the larger lumps.

Erosion is just one side of a glacier's powers of landscaping and it is a truism to state that what the glacier removes so too will it deposit, usually in another place. Just as there were two separate faces of glacial erosion, abrasion and plucking, so too are there two faces of deposition, dumping and sorting, or to give them their proper names, glacial and fluvial. In the former, material held in the ice is simply dumped as the glacier melts or moves on. The resulting deposits are unstratified, which means their components are irregular in shape and are all mixed up higgledy piggledy and are called till. Till comes in the form of more or less uniform sheets called ground moraines and they cover large areas. The largest ground moraines are formed where the glaciers have finally melted, dumping a layer of both sub and supra glacial till, material carried both below, in and above the ice. The surface of ground moraines may be fluted with ridges that run parallel to the direction of movement of the glacier. Till may also be dumped along the margins of the glacier or at the foot of the glacier as it melts; such ridges or mounds may be

Snowy Owl with Sea Plantain and Bear Berry

PLATE 10

very large and are called either lateral or terminal moraines. The third expression of glacial dumping is where the till forms elliptical hills called drumlins; their exact method of formation is still somewhat obscure because of their complex internal makeup but their presence in a glacial landscape is often unmistakable.

The problem is that throughout the life of a glacier, melting is going on at certain times of the year and meltwaters are being produced. Thus it is often difficult to sort out where dumping stops and sorting starts. However a close look at the deposits themselves should leave little doubt in the mind. Sorting takes place as the glacial debris is carried along by water and therefore some of the fragments will be rounded by the action of flowing water and the deposits will be stratified, that is sorted into layers each consisting of different sized particles. The faster the flow the larger are the particles that can be carried along; as the flow slackens the large ones are deposited first, then successively smaller and smaller ones fall to the bed of the meltwater stream until only the rock flour is left in milky suspension. If the flow slackens sufficiently then even the rock flour will be deposited to form what will eventually be called clay. The sorted deposits are laid down in the following order: boulders, pebbles, gravels, sands, silts, and clays, and the landforms they produce also come in many shapes and sizes.

Eskers, which in their pure form consist of long sinuous ridges, are formed in tunnels beneath or within the ice, or in channels eroded by meltwaters in the surface of the ice. Mounds of sorted material formed from deposits accumulating in lakes or on pockets of water under the ice are called kames and if they were formed in lakes situated along the margin of the glacier they form kame terraces. Often numerous kames are found together, producing an undulating topography, the depressions between being termed kettles. Some of the kettle holes may be very deep and steep sided, especially where dead ice was pushed down into the underlying drift by opposing pressures set up within the glacier.

The largest of the meltwater deposits are, however, found exactly where you would expect them to be, located at the foot of a receding glacier. As the glacier disappears or, to give it its proper name, ablates, the meltwaters are charged with the products of erosion which they dump in the form of great fans or sandar. If, as often happens, the terminal moraines left by dumping from the glacier dam the valley then the meltwaters will form a lake and the successive years' melt will slowly fill the lake with sediment. The annual pulse of meltwaters into such a lake produces a series of micro strata. In their finest form these are called varves and can be used not only to date the sediments but also to produce an accurate picture of the successive annual climates at the time they were formed, for their thickness and composition mirrors the annual rate of melt.

The study of a series of glacial deposits is never simple because of the complexities of strata laid down during the genesis, life and ablation of the glacier, the complexities of advance and retreat and the fact that, what dumping may build, meltwaters may re-sort. However, the wealth of information which may be gained from such a survey makes it all well worthwhile.

The exciting thing is that one doesn't have to be either a geological or glaciological expert to get to know the story in the rocks and the message in the landscape; with a little knowledge and a willingness to learn it is possible to begin to read the past of your own landscape in detail, interpreting the passage of The Great Seasons right in your own backyard. Well, almost.

So it was that the landscape of the Upper Dale as we know it today came into existence and examples of all the features mentioned above were exposed to view or came into being as the glaciers diminished in size. The final 'permanent' ice disappeared from the Dale around 12,000 years ago and it melted exactly where it had first formed more than 6,000 years before in the highest lateral valleys. Well not exactly, for erosion had taken its toll, reshaping these highest sections of catchment of the river into the form of corries. The scene was thus set for the Great Spring to come, but first, what do we know of life throughout that Great Winter?

The edge of the Ice Sheet. Arctic Skua, young Herring Gull

PLATE 11

Life during the Great Winter

THE ICE SHEETS stretched far south of the Dale leaving their terminal moraines, unstable evidence of their presence, in a ragged line from Bristol to York and out across the northern world. They eroded and reshaped much of what we now call Europe, Russia, the United States and all of Canada, producing their own distinctive landforms which are part and parcel of our everyday lives. If you don't believe that the countryside in which you live was once beneath ice, go take a look, learn to read the scratchings on the rock, take countenance of the U shape of the valleys and those weird mounds in which today we quarry for gravel, sand and the other remnants of pre-glacial landscapes. When you have found them, stand and think, not only of the crushing weight of white death which once stood on that place but of the way in which the Great Winter cleansed and prepared the ground for future events, which include your own way of life. Think, too, that throughout that span of time life went on both around the margin and above the sheet of death and it was those hardy plants and animals that survived closest to the ice that would have acted as the pioneers in the recolonisation of the Great Spring.

As in all endeavours of science, or come to that in all endeavours of thinking Man, there is argument which, like an ice sheet, can be both a destructive and constructive force. One such discursive disagreement revolves around the presence or absence of nunataks within the Upper Dale. A nunatak is a mountain peak which stood bare of ice throughout part or all of the period of glaciation. If a nunatak did manage to keep its head above the solid water then it could have formed a refuge for certain hardy types of life throughout the Great Winter. Such small populations of living organisms would have then formed inocula for the process of recolonisation when the climate ameliorated and the Great Spring began in earnest. Within the Upper Dale good candidates for these tough and elevated positions of survival would be Yellow Marsh Saxifrage, Stiff Sedge and Alpine Foxtail, all of which still grow not far from the highest peaks on the western boundary which would have stood the best chance of nunatak status.

I don't think anyone has ever come to blows concerning the presence of nunataks in the Dale, but the argument has caused many eminent men much discomfort as they tramped the highest fells seeking out the evidence for or against their existence. Unsorted tills undoubtedly dumped *in situ* and ice plucking of the rocks show that even the highest peaks were overridden by ice. Yet lower on the flanks those same peaks bear the undoubted mark of ice worked patterns including some superb stone stripes. Such patterning is only produced in permafrosted ground beyond the compass of surface ice and so would indicate that the peaks were clear, each one a centre of survival.

To understand the process and the complexities of interpretation one has to travel to the present margin of the northern ice cap in the high arctic. In comparison the contemporary ice is a shadow of its former self, yet stretching far to the south of its present terminal moraines is a sheet of hidden ice, the zone of permafrost in which sub-surface temperature remains below the freezing point of water throughout the year.

Each winter the permafrosted ground is frozen solid to the surface and yet the warmth of each new spring melts the hidden ice to produce an active layer charged with minerals which are the final products of erosion. As summer advances this layer in which all the action happens, deepens and more water wells up to the surface, but not for long, for as autumn comes the refreeze starts and soon the whole land surface is solid ice once more; the active layer is once again quiescent. This annual cycle of refreeze and thaw motivates the patterning. The expansion and contraction of the water as it changes from liquid to solid state and back again pushes and heaves at the glacial debris, sorting the large from the small, forming them into neat patterns or moving them downslope. This process of traction is

POLAR BEAR

PLATE 12

called solifluction and, in the main, the angle of slope of the terrain controls its end result.

On flatter ground frost heaves the sediments up into mounds within which differential ice formation pushes on the larger lumps ejecting them from the pile, to produce a neat circle of stones arranged round a pile of finer debris. Each then becomes a focal centre for the development of vegetation, the first plants often growing around the margins in the shelter of the ramparts of tilted stones. The finer material in the centre remains quite bare until frost heaving is no more, a thing of a colder past, when plants move in and find a sure foothold and easy rooting space.

In other places the surface may dry and crack, producing neat polygons. Meltwater then flows into these cracks and on re-freezing forms ice wedges that push deep into the terrain. Each one becomes a solid edge against which material is piled up, raising the edges of the polygons and again providing shelter for the growth of the first plants.

Where the surface slope is in excess of five percent such polygons become distorted, elongated downslope, eventually forming stripes of large stones separated by finer debris, which run almost parallel downhill for many tens of metres. On the steepest slopes solifluction simply loosens material, sending it on its own way to tumble down to rest in the form of screes made up of shattered blocks of rock called regolith. In this way new surfaces are continually laid bare, and frost erosion continues, lowering the surface, adding to the constant rain of regolith. Erosion by frost shatter may continue long after the permafrost has gone away.

This is one reason why all good gardeners lay bare the earth in winter to allow the frosts to penetrate and help break up the soil. After a good hard frost it is possible, with the aid of nothing more sophisticated than a pen knife, to discover ice lenses at work in your kitchen garden; and needle ice is there at work on every cold and frosty morning.

Ground with regularised patterns, circles, stripes and polygons is in the main produced while permafrost is still present in the ground. So it might be argued that the flanks of the High Fells, where the imprint of such patterns may still be seen today, stood clear of any ice burden, a nunatak, a refugium for life throughout the Great Winter. There is, however, no reason to doubt that those same patterns could have been etched upon its surface as the Great Spring began to break and the High Fells lost their burden of solid water. Although today I feel the consensus of the evidence suggests that the whole of the Dale disappeared beneath the blanket of white death, perhaps it is more constructive to leave some doubt to salt the academic tables of the future.

There is, however, good evidence for the fact that the ice was never all that thick over the top of the peaks and that, early in the Great Spring, downwash soon cleared the upper slopes of ice, opening them up to frost shatter and solidification but leaving the valley itself filled with its own glacier once again.

In my travels I have spent as much time as possible in the high arctic realms of Canada where it is possible to gain the experience of living in the environs of the ice cap. A warm summer's day on Ellesmere Island, high on the arctic ice, has a magic all of its own. The only other experience I have had which can compare is underwater on a cloudless tropical day floating just above a wide expanse of pure white coral sand. 'Mind blowing' is much too brash a phrase yet in it you can sense the effect of immense white bowed horizons set against purest azure. An albedo of cold warmth rebounding back and forth, refracted into rainbow facets which dance in the interplay of light and shade. Stand still and all is silence apart from a background rustle of melting ice, the merest whisper as ice crystals implode, giving up the energy that has remained latent for so long, and take up their smaller volume. Liquid water which will remain in that state until the sun sinks down below the ice horizon, when each cubic centimetre of water will once again take on its new enlarged dimension, setting up a jingle which runs across the surface of the glacier slightly in the wake of the advancing shadow.

When water was created it would seem that it was made to disobey the rules, for when most substances change their state from liquid into solid they increase in density, their molecules packing closer. Not so with water. At around 4°C the molecules in liquid water are as closely packed as they will ever be and water at this temperature will sink. Below this temperature ice crystals begin to form and as they do the water molecules will be pushed further apart and the resultant solid water will be less dense and will thus float on the surface. If it were not for this one quirk in the nature of water there would be no life on earth, for at the advent of each new winter the ice would sink and the lakes and seas and oceans would soon be frozen solid and the world would remain locked in a winter of solid ice.

Water, however, does not disobey the rules. It cannot, for they are themselves the acts of creation containing as they do the behaviour of water in all its states, gas, liquid, solid and structured into the panoply of life.

A walk across the surface of the ice breaks the silence, as your feet stir the ice water mix, sending the packs of ice candles toppling until they come to rest tight packed against each other. It seems like sacrilege to move, for movement will despoil the scene of iced perfection, but there is so much to discover and in the knowledge that at the setting of the sun all will form again, obliterating the evidence of your brief presence, you can press on.

In places where the ice surface slopes gently away from the direct rays of the sun the surface does not melt except in areas where impurities occur upon the icy scene. During the life of any glacier dust will blow across its surface to become trapped by the next fall of snow. The dust, being darker than the snow itself, will absorb more heat and this dirty ice will melt more rapidly, creating depressions into which the dust will concentrate. Circular dust bowls are thus formed and continue to absorb more heat than the pristine white snow round about. The depressions deepen, studding the surface of the glacier with deep vertically sided pits each almost filled with pure blue water and capped each night with a gossamer of ice fretted and worked by the stresses of crystallisation. During the day, depending on the amount of cloud and thus the amount of surface radiation, these caps may completely disappear or stay like the lid of a German beer mug hinged to the shadiest point of its circumference. Some may be more than a metre across and as deep, others but a few centimetres in each of its dimensions, yet each bears in its depth the measure of its formation, a layer of dark dust full of minerals and packed with future promise.

I once sat afar and watched in awe as a Polar Bear tracked across the ice inspecting a series of such holes, perhaps to drink or perhaps thinking that they were the blow holes of his favourite prey, the seal, not knowing that he was far inland. I wonder if these magnificent creatures, whose furry padded feet give traction and protection on the ice, came to the Upper Dale in the scant summer days towards the end of the Great Winter? They must have lived that far south, feeding in the iced fringed seas and if, like their modern counterparts, they did make excursions far inland they could have stood above the ramparts of the Dale and gazed in hungry hope into such holes. However, if they did they would have been but short excursions, like mine upon an arctic summer's day.

The horizon changes, a haze of salmon pink. I check my watch. It's nowhere near the time when the sinking sun will flush the whole scene with the same hues. Yet all around it is distinctly pink. I turn, look back; the middle and far distance are as white as any housewife of our modern world could want, but there in my footprints is the answer — biological! No, not some new washing powder, but the only imprint of permanent life upon the surface of ice.

Each footprint I have left is picked out in deeper red, in places almost livid brown, the unmistakable mark of the Snow Algae. This tiny plant-like animal (for it can both fix the energy of the sun, turning it into sugar, and actively move, swimming in the water between the melting ice crystals) has not sprung to life in my footprints; they are there all around, a living pinkwash suffused throughout the part melted crust. My weight has done the trick, compressing the melt, concentrating both the culture and the colour.

To investigate their world our human eyes must use a compound microscope, for only its power of resolution can reveal their tiny landscapes, disclosing wide open spaces, angular horizons, a jumbled hall of mirrors reflecting the light which makes their very special way of living possible.

Each tiny animal has a single eye, a red spot of pigment with which they see

Arctic Hare and red snow.

PLATE 13

blindly, for it only allows them to discern between light and dark. They are thus able to orientate cyclops-wise to the most advantageous environment where there is sufficient light to activate photosynthesis, yet not enough to destroy the pigment chlorophyll which is, especially in its younger state, broken down by too much light, and in such a white world light can be problem.

I can only wonder, do those lustrous crystal halls echo to the beat of their flagella which propel them through the water, and is the pigment which suffuses the rest of their body there as an energy store or to protect the green pigment from the devastating effects of too much light?

The life of *Chlamydomonas nivea* is very short; but a few hectic days during which they must grow, live, move, have their being, mate and reproduce, before the onset of the next winter. The product of their love's labours is, however, not lost, for the next generation is wrapped within a thick coat, a resting spore which can withstand the onslaught of the hardest normal winter. Multiplication within the spore produces a mass of new cells called a palmella, new potential which may dry and be blown away by the wind to colonise another fall of snow. Like all other forms of life the secret of *Chlamydomanas nivea*'s annual success depends on the resources of their chosen plot. An important part of the resources of their particular habitat are the minerals in the dust that settled ages since upon the virgin snow and will one day speed its destruction.

In nature nothing goes to waste. The Snow Algae are but one part of a living system, the power houses which fix the energy, chanelling it to good use and when they die their bodies will not taint the perfection of the snow but form the food for fungi and bacteria which share their frigid environment, cleaning up and recycling the minerals into the bargain. This is no case of undertable crumbling charity, but a microcosm in which every organism has a role to play and, what is more, that role is always played to perfection.

Apart from these tiny organisms all other forms of life above the ice are but transient visitors who, like me, come to 'blow their minds' upon the freshness of another arctic day before they hurry down to warmer lands.

What birds could have made short excursions above the ice? Arctic Skua, Snow Bunting, Barnacle Geese? Who knows? It is but guesswork, and yet I feel sure that Arctic Tern flared their flight feathers against the summer sun as they flew along those cliffs of ice, rejoicing perhaps in the knowledge that in those clear days of the Great Winter their annual migration, which still spans the globe linking the sunshine of the poles, was compressed in places to almost half its normal warmer distance. If they did alight upon the ice to rest, I wonder, was the stamp of eight tiny spreading claws sufficient to squeeze the redness in the snow to visual recognition?

Life Around the Margin of the Ice

The transient nature of life upon the ice means that there is no record of its presence to be read by man or woman, however learned. The same is fortunately not true of the life which existed around the margins of the glacier, for there, in certain favoured spots, life left its imprint deep and clear; an imprint that would be brought to the light of twentieth century endeavour.

There is now little doubt that whatever caused the ice sheets to develop also deflected the more southerly climatic zones, permafrost tundra, and non-equatorial forest, down towards the equator. The latter appears to have lost most ground, its overall range having been reduced to less than half its former status. This was at least in part compensated for by the shrinking of the world's oceans.

It has been 'guesstimated' that at the height of the glaciation more than 80,000 billion tonnes of water had been effectively removed from the oceans and dumped upon the land. Now although this represents only about 6% of the world's total water supply, the shift in its emphasis was sufficient to warp the earth's crust, depressing all land areas bearing glaciers and make the world's sea level go down by as much as 100 metres. This meant that many of the world's oceans were smaller than they are today and shallow stretches, like the English Channel, much of the North Sea, and the Bering Straits, became dry land, giving direct access for anything or anybody who wanted to make the frigid crossing. No

ARCTIC TERNS

PLATE 14

water barrier, no ports, no customs or immigration; a nice summer stroll between the White Cliffs of Dover and Cap Gris Nez, Stranraer and Larne and indeed between Russia and America, and there were men in plenty who pioneered the routes.

They came and went in search of rich pickings: Mammoth, Woolly Rhinoceros, Bison, Reindeer, Elk, to name a few that have since disappeared, at least in part at the hand of Man, from the face of Britain, and in the case of the first two from the face of the *world*. Large herds of these animals found sufficient sustenance from the scant vegetation of the tundra zone, themselves migrating north and south with the pulse of the annual seasons. The herbivores were preyed upon by Wolf, Arctic Fox, and, of course, by Man himself, while omnivores like the Brown Bears, fed on the plenty of the summer runs of Salmon, Sea Trout, and their like, moved up the southern rivers and gorged themselves upon the berries of the tundra before they went into hibernation.

The records of this wealth of life come, in the main, from organic deposits trapped in gravels which were laid down by meltwater streams flowing across open ground. One can imagine the annual collection of natural bric-à-brac: bones, twigs, branches, teeth, brought down by the first rush of melt, to be deposited, still frozen, in some backwater. There they would be rapidly covered and sealed in by gravel, sand and silt from the oxygen-rich air above and hence from the main process of decay. Where conditions for preservation were best the organic record preserved even the more delicate natural objects which came its way, the bodies, legs, and even wings of insects, and the leaves of tundra plants. What is more the pollen and spores of many of the plants, though small in comparison to their other parts, are tough and very resistant to decay. Each one has an outer coat of cutin which is there to protect the living information held within, information concerning the form and function of the next generation which will be revealed only when the spore germinates or after pollination when the seed develops to fruitation. However, in the right hands and with the powerful techniques of modern science, it is possible in many cases to identify the plant in question from the sculpturing of the outer protective case of its spores or pollen grains, thus, as it were, usurping the rights of the new generation to reveal the secrets of their lineage. Such study forms the basis of the exacting discipline of palynology (pollen analysis) which has produced and still is producing much of the information about the cycle of The Great Seasons and life around the margins of the ice.

Detailed study of both the macro and microscopic remains reveals a picture of an open landscape, the vegetation of which was rich in sedges and herbs with brightly coloured flowers pollinated by a buzzing array of insects and almost devoid of trees of any sort. Plants like the Arctic Poppy, Snowy and Woolly Cinquefoil, Arctic Buttercup, and Two-flowered Sandwort, to name but a few, grew in profusion in these southern tundras, along with Arctic Willow and Dwarf Birch and Dwarf Juniper, which, although trees by name are small shrubs by nature and rarely grow as high as 20 cms. One of the reasons for the lack of tall trees would have been instability of the ground, underlain, at least in places, by permafrost; the surface layers actively heaved and sorted would soon topple anything which grew up even to mere sapling size.

Having wandered over the tundras of Canada's high arctic and watched herds of Musk Oxen and Caribou graze on the bounty of sedge meadow and fell field, tormented by myriad blood-sucking insects, it is quite easy to imagine the scene painted in scientific detail in the reports. Yet until we understand what it was that caused The Great Seasons to happen we will never know for sure exactly what the environmental conditions prevailing at the time were like.

Close to the poles the march of the seasons is hammered home each year as the sun disappears below the horizon and remains out of sight throughout the winter night. The length of time for which it remains thus hidden, and during which the landscapes receive no direct radiant heat, depends on the exact latitude. Likewise for a number of days each summer the sun never sets and the patient observer may sit around the clock and watch the sun make a complete circle, its fiery life-giving orb dipping down to wash the northern ice rim in a golden redness, a colour I have seen nowhere else on earth. The arctic and antarctic circles are the official marks

WOOLLY RHINOCEROS

PLATE 15

beyond which the inhabitants need never fear the daily cycle of cold dark nor will they experience the pleasures of the midnight sun in the summer.

If the ice age was caused by the tilting of the earth — and remember it is no more than mere hypothesis — the winter night and midnight sun could have been experienced far to the south of what is now the arctic circle, and the environment would have really deserved the term arctic. If the ice age was due to some other cause then the environments of the periglacial fringe would have been much more like that of the present alps, where even in the depths of winter the days are warmed by the sun.

To date the evidence we have comes from a scatter of sites brought to light by chance excavation. There must be many more waiting to be found amongst the rolling vistas of the English southern scene and far beyond across the world that once lay on the fringe of the Ice Cap. When more have come to the light of modern investigation we may be able to stop our educated guesswork and perhaps make some decision both about the exact effects and their causation.

But let us recap on the evidence at hand: we know that some 120,000 years ago the climate of the world began to change and ice sheets began to form. All non-mobile forms of life were killed, fixed to their own appointed spots, while those which could migrate moved south as far as they were able. The road, however, was not all clear, for the potential of those more southern climes was already filled to overflowing with their own communities of plants and animals, some of which, under the stress of climatic change, were also forced to migrate to the still warmer south. There were, of course, some compensations, for as the sea level dropped new land was opened up ready for colonisation.

In the main it may be said that as far as life was concerned the ice sheets were a force of destruction and change on a massive scale and at its zenith tundra vegetation was found at least as far south as 45°, the latitude of Bordeaux, more than 600 kilometres beyond its most southerly extent today.

We also know that throughout the Great Winter ice action was preparing the ground, shaping new landscapes, and that the plant and animal communities of the broad tundra fringe would provide the pioneers for recolonisation when spring returned some 12,000 years before the present day, releasing liquid water to perform a multitude of jobs.

Young Heron

Minnow   Water Flea   Mayfly   Common Frog   Mud Sedge

PLATE 16

The Great Spring
Books of living History

IQUID water moves across a landscape dissipating its potential as it flows, until some feature checks its progress, so forming ponds and lakes. Any body of moving water soon becomes a highway and a habitat for life, the rule of the watery road being, the faster the flow the less life will be present. The problem is simply hanging on or at least maintaining station, and both require the utilisation of energy. There is, however, one consolation for the members of such lotic environments; usually the faster the flow the more oxygen is present, dissolved in the water, aiding them in the maintenance of the necessary flow of energy. Water may be a good solvent and an active agent of erosion but oxygen dissolves in and diffuses through water at what amounts to an insufficient rate to keep an active living system going. Mass flow, especially when it is turbulent, as when the water falls over rock ledges and rapids, speeds both processes and keeps the water charged with oxygen, almost as 'fresh' as the air which is constantly stirred in. As soon as the rate of flow slackens, oxygen supply problems rear their ugly heads and aquatic life begins to gasp for air. Remember what happens when the bubbler in your aquarium stops bubbling.

Thus it is that in the lower, slower reaches of streams and rivers, and especially in backwaters, lakes and ponds, although life is on a firmer footing it faces greater and greater problems of oxygen supply.

Every body of slowly flowing water thus becomes a habitat in which living systems rapidly develop. The sodden landscapes newly exposed by the melting of ice were replete with such aquatic microcosms in kettle holes, moraine dammed lakes, permafrost patterns, in fact the whole textbook-full of glacial phenomena overflowing with meltwater.

At first water plants grew in profusion, their remains falling to the bottom, their matted stems trapping silt and other organic debris. While sufficient oxygen was still present much of the plant and animal remains were broken down by fungi, bacteria and a host of sub-aquatic garbage-decomposers. Only the toughest parts of the plants, seed coats, fruits, spores and pollen, remained in a matrix of fine silt and organic debris, lake muds which we call *gytta*. Semi-aquatic plants, such as sedges, rushes, horsetails and mosses grew out from their margins into pools shallowed by the remains of communities which had grown there before. These emergent plants in part overcame the problems of a dearth of oxygen because the bulk of their green shoots pushed up into the fresh air above the stagnant water. However, down below in the tangle of roots and rhizomes and the annual fall of leaves and shoots, the habitat soured, became devoid of oxygen and the process of decay came almost to a stop. As it did the record held in sub-fossil form became more perfect and more detailed as peat began to form.

Each accruing peat deposit, however large or small, thus became a chronicler, a diarist, taking down the record not only of what took place upon that stagnant spot but, thanks to the annual rain of pollen spores and dust, of what took place within its catchment and beyond. These peat deposits were thus living history books that would record the progress of The Great Seasons as the potential of 20% of all the world's land surface was opened up to life once more.

Of all the organisms present in that periglacial fringe one above all is the epitome of this part of the story. *Lemmus lemmus* is very much at home in the conditions of the tundra, feeding on lichens and mosses through which they make tunnels where they continue to live and feed throughout the winter, safe under a covering of snow. They are in fact safer during the winter period than they are in summer when a greater diversity of predators cull the weaker members from their swelling ranks. Buzzards, Raven, Chough, Arctic Skua and Snowy Owl join Stoat and Weasel in the summer feasting.

A female Lemming may bring forth several litters in each year, the period of gestation being as little as twenty days, after which three to nine young are leased

Lemmings with Mountain Avens and Mountain Sedge

PLATE 17

upon the world. In the absence of the bulk of flying predators the winter broods have a better chance of survival and, if the conditions are exactly right, spring will find the potential of certain areas full to overflowing. It would appear that when the population in any one area gets too large the stress of cohabitation and diminishing resources tips some quasi-social balance and the Lemmings begin their explosive migration. They do not rush headlong down towards the sea to commit mass lemmicide; they move across the tundra radiating in search of food and fresh ground in which to set up their own individual way of living. In normal years such haphazard migration must in many cases end in catastrophe and hence the stories of mass death at river crossings and the margin of the sea. Such behaviour is, however, part and parcel of their mechanism for survival, and during the Great Spring it may well have been *Lemmus lemmus* and perhaps a number of the other twelve species of these rodents that were the first mammals to range across the new landscapes. As they made their explosive way north, in each good lemming year, so they would have drawn their predators with them, potential building on potential. The success of one was both determined and controlled by the success of the other, each one being part of a dependent complex web of existence.

However, in the absence of green plants to fix the energy of the sun, the presence of animals of any sort would be but a transient phenomenon dependent upon the amount of energy they could store or carry with them. It is thus clear that green plants must have been the first harbingers of the Great Spring to come, but how did they make the journey?

For almost 5,000 years southerly winds had carried the propagules of tundra life, spores, light seeds and insects bursting with eggs up to perish on the barren slopes of the ice cap. Now these same air streams carried the promise of new generations to fruition, vitalising those horizons which had for so long lain beneath the ice in the dark grip of the Great Winter night.

So it was that the first green fingers of spring moved north, eventually to touch the flanks of the Upper Dale which remained festooned with ice as long as any other part of England. The frontiers were open: the land was up for grabs and the Great North Road was as wide as the margin of melting ice, a road that spanned the northern hemisphere.

The Great North Road

When I was first taken up from the warmer south towards the cooler north of Britain I travelled the Great North Road, or, to give it its less romantic name, the A1 motor route. It was then a highway which had, in the process of maturation, thrown off the purposeful straight lines of youth that sped the Roman Legions on their way, to meander, a lazy ribbon of tarmacadam bejewelled with cat's eyes, linking the towns and villages in its path.

My first real memory of the Great North Road was towards the end of the winter of 1948-49 when conditions were as bad as they had been within recorded memory and people said another ice age was on the way. We did not get far. I stuck in a mass of driven snow packed firm by all the other homebound traffic pouring north from the metropolis. One could only wonder what conditions would be like further to the north and think back to the Great Winter.

What then were the conditions really like across the broad swathe of Britain as this shrinking peninsula emerged from beneath the blanket of white death? How was it that as the Great Winter ablated into spring the plants and animals were able to make their various ways up from the refugia south of the line of glacial advance? The answer is that the potential was there and it had to happen.

It is so easy for us humans to put our own limitations onto other members of the living kingdom and even onto our own forbears. Conditions which present our vehicle-bound age with a lot of travel trouble, bringing communications to a halt, may well be ideal for other animals and plants.

As the ice melted the great sweep of land that would one day be England, complete with pools of melting ice, lakes, eskers, moraines, drumlins, ice-smashed rock and finest glacial clay, was gradually revealed. This was not only

The end of the Ice Age. Gadwall pair with eggs and duckling

PLATE 18

their land of opportunity but a broad highway of migration for the overflowing population of plants and animals crowding the arctic fringe.

It is the wont of all living organisms that each generation produces a surfeit of offspring, far too many to thrive on the resources of the land. Only certain insects and all mammals appear to have evolved beyond the limits of this rule, their social order or parental care apparently reducing the wastage of new life. Yet, even in the case of the most advanced society each new generation is potentially larger than the last whatever are the limitations of resource. Think no further than the present dilemma of mankind.

Each population of every species plays its own generation game in which offspring compete with parents, and, what is more, the stakes are as high as they can be; survival in the struggle for existence. The outcome is that, in the main, the fittest survive and the less fit go to the wall. Thus it is that this apparent waste of overproduction is a key feature, not only of survival but of the mechanism from which advances stem, for this is the stuff of natural selection.

Please, even if you reject in its entirety the theory of evolution and hold fast to the belief that God created each and every species, remember that annual over-production and the fact that only the fittest survive is part of the created plan. Natural selection is a fact of life; under the guise of evolution it is a creative force; within the conditions of special creation it keeps each element of creation honed to perfection.

Regarded in this way each population of each organism is a potential explosion contained within its present bounds, there to be unleashed when the opportunity arises and at the end of the ice age there was opportunity on a massive scale.

Hooray for the M1

Perhaps it is only fitting that one of the living history books which has yielded information of the advent of the local spring was found during excavations to re-route the Great North Road along a wider, straighter course turning it into a motorway — M1. The site in question, all that remained of the arm of a once extensive lake, is not far from one of the oldest railways in the world and almost in the centre of the lower river basin, some 75 metres above sea level. To provide firm footings for the new road it was decided to excavate the least consolidated deposits from the basin. Difficulties in the stabilisation of the substrata led to advice being sought from the nearby University and thus the site was brought to the attention of botanists and geographers.

One of the most fascinating aspects of the site was a deposit, in places nearly one metre thick, which consisted of three mosses which were in such an excellent state of preservation that they could be identified without difficulty. They were *Calliergon giganteum*, *Paludella squarrosa*, and *Homalothecium nitens*. My apologies for the Latin names; they are however necessary because the majority of mosses, and come to that most of the less conspicuous plants, have no common names which are widely recognised. All three are common plants of contemporary tundra and all three occupy a very special place in the succession of vegetation which colonises and gradually fills such sites with peat. Below the well preserved mat of sub-fossil mosses was a thin layer of lake muds and marl, below which was a smearing of clay, overlying and sealing glacial till.

The picture was thus of a water filled depression supporting communities of Stoneworts and Alternate Leaved Pondweed which provided grazing for two sorts of snail, Common Valve and Ramshorn, and shelter for two bivalves, Freshwater Mussel and Pea Mussel. The abundance of snail shells shows that the water was rich in lime (calcium carbonates and bicarbonates) and the very bad state of preservation of the organic material present indicates well oxygenated water. The presence of both Common Reed and Great Pond Horsetail in the basal deposits emphasises the shallowness of the lake at that point and suggests a rapid rate of infill and succession towards a new type of vegetation dominated by a mat of mosses supported, at least in part, by the matted rhizomes of Bog Bean and Marsh Cinquefoil.

At first the Spear Moss (*Calliergon giganteum*), a plant which still thrives in the Upper Dale around springs of calcareous water, dominated the mat, its

New Landscapes, the Lesser White Fronted Goose, Thrift and Alpine Bistort.

PLATE 19

place eventually being taken by *Paludella squarrosa* which remained in its dominant role for a long time, the mat rising and falling with any changes in water level.

The underside of such a floating mat would thus always be in contact with the lime rich ground water which would itself move by capillarity up through the mat, keeping the living green moss shoots well supplied even in the driest weather. However, when it rained exactly the opposite would happen; the almost pure rainwater percolating down through the moss mat would carry any soluble minerals with it.

All living things, and that includes you and me as well as *Paludella squarrosa*, produce acidic waste as by-products of their living chemistry. In such a floating mat these would be washed away to be neutralised by the lime rich water beneath, the whole mat coming into a sort of shifting equilibrium, alkaline at the bottom and more acid at the top, the exact balance depending on the weather conditions prevailing at the time.

As the arm of the lake filled and the floating mat thickened, settled and consolidated, the surface conditions would shift more and more towards the acidic side. That such a change did take place is indicated by the gradual replacement of *Paludella* by two sorts of bog moss, namely *Sphagnum teres* and *Sphagnum subnitens*. As their name suggests the bog mosses are denizens of bogs, places in which the water is usually acid and contains little or no dissolved lime and other minerals as it is derived mainly from the rain falling directly on the surface of the vegetation. Both the species in question, though able to tolerate mineral rich flowing ground water, do mark the change over to more acid conditions for they were in turn colonised by Cross Leaved Heath and Heather both of which are calcifuge, which means lime hating.

All this discussion of chemical change may seem out of place. However it must be pointed out that these changes recorded in the living history book were just part of similar changes which were going on apace across the face of the Dale, in fact along the whole of the periglacial fringe. The newly bared land surfaces charged with all the minerals released from the bedrock by glacial action were now opened up to the leaching power of the rain. Soluble minerals like calcium and magnesium bicarbonate would be carried away to enrich the ground water leaving the surface of the developing soil poorer in minerals and with a tendency to become more acid.

The presence of permafrost must have slowed the process of change as the annual re-freeze and thaw stirred the deeper horizons, recharging the active layer with minerals which had been washed down the developing soil profile. The presence of the layer of ground ice itself acted as a barrier to deep leaching. However once the environment had warmed up sufficiently to melt the last vestiges of ground ice, leaching would become a major process effecting pedogenesis, the process by which a non-living mineral material becomes a living soil.

Apart from the loss of lime and the development of soil acidity one other major problem relates to increased leaching and that is the supply of nitrogen to the developing vegetation. Nitrogen is one of the big three plant nutrients, the other two being potassium and phosphorous, and all farmers and gardeners know to their cost that they must add them to the soil to ensure a good crop each year.

Although nitrogen is abundant in the atmosphere it is there in elemental form, just plain nitrogen, which is inert and as such is not available to plant growth. Nitrogen is also not very common in the rocks which make up the crust of the earth and from which soil is made, and when it is present it is usually in the form of nitrates, all of which are notoriously soluble and hence would be immediately lost by leaching. Thus the development of anything but a very scant vegetation would be impossible or at least would be strictly limited by the lack of nitrogen. Nature has, however, overcome this problem through a chemical process called nitrogen fixation. Certain very simple organisms, bacteria and cyanobacteria, have the ability to fix atmospheric nitrogen, turning it into nitrate. Although these organisms are in the main very small they are present in all types of natural vegetation and they are of immense importance.

The nitrogen fixing organisms reproduce by means of simple cell division and may under adverse conditions produce resting spores. These are very small and light and hence may be readily blown about, inoculating new ground. It would thus be fair to say that the first important step in soil formation is always inoculation with nitrogen fixers. It also seems safe to suggest that the first flush of the Great Spring was blue-green in colour, for this is the colour of the pigments which allow the cyanobacteria (blue-green algae as they used to be called) to fix the energy of the sun, turning it into sugar through the process of photosynthesis. Part of this energy would then be used for growth and part for the fixation of nitrogen, enriching the habitat with this all-important nutrient. Once active nitrogen fixation had been established the whole process of colonisation would have been almost self generating, potential building on potential, the exact course depending on what dropped in in the form of propagatable material.

On a number of trips up to the High Arctic I have taken all the paraphernalia with which to test for the presence of nitrogen fixation. In all cases I have obtained positive results even from the barest of soils close to the permanent ice. There was however no real need to take the gear, for a careful look revealed the presence of cyanobacteria on the surface, often as jelly-like blobs and sheets but also minute filaments filling the spaces between the mineral particles. To see their full beauty it is necessary to take a look under a compound microscope when it may be seen that even the largest sheets, which may be as much as a metre across, are made up of chains of elongate cells with larger more robust looking cells at irregular intervals. Each chain or filament looks not unlike a delicate necklace of precious stones, the larger ones, which are called heterocysts, being the 'gems' of the piece. Indeed they are, for although it has not been definitely proven that they are the seat of nitrogen fixation it is known that of all the cyanobacteria those with heterocysts have the best capabilities of performing the important function of fertilising the fields of the world.

In my travels around the world I have been very careful to search for modern plant communities which resemble the sub-fossil assemblages known from the living history books of the Dale. The nearest I have ever found to the moss mats of

of the reputed 10,000 lakes are in the final stages of infill by peat-forming
the motorway deposits were in the Mazurian Lake District of Poland where many
vegetation.

The Mazurian Lake District is in fact a large area of glacial debris left by the
melting of the continental ice sheet, its surface moulded by deposition and carved
by subsequent erosion into a jumble of lakes of all shapes and sizes. Many of these
have been little disturbed by Man right up to this day. *Paludella squarrosa* is not an
uncommon plant in some of the lake and stream systems and it always occupies a
similar position which is often very treacherous to approach, a semi-stable mat
developed over flowing calcareous water often close alongside an almost hidden
streamlet or indeed all that remains of the lake itself.

The following list gives the floristic makeup of the 'moss mat' communities of the
Wilderness of Pisz in the Mazurian Lake District of Poland where the process of primary
succession started during the Great Spring still continues to this day. Their presence in the
living history book which now lies beneath the A1M motorway is shown as follows: M
denotes macroscopic remains identifiable to the species level; P, pollen grains identifiable
to the species; PG pollen identifiable to the generic; and PF pollen identifiable to the
family level only.

VASCULAR PLANTS

Bog Bean, M; Bottle Sedge, M; Broad Leaved Cotton Grass, PF; Bullrush, M; *Carex chordorhizza*, PF; *Carex diandra*, PF; Common Meadow Rue, P; Carnation Grass, M; Common Reed, M; Common Black Sedge, PF; Common Spike Rush, PF; Cranberry, M; Dioecious Sedge, PF; Fen Bedstraw, PF; Greater Duckweed, King Cup, M; Large Bird's Foot Trefoil, Lesser Bladderwort, Lesser Duckweed, Lesser Reed Mace, M; Long Leaved Sundew, Many Headed Cotton Grass, PF; Marsh Bedstraw, PF; Marsh Cinquefoil, M; Marsh Fern, M; Marsh Helleborine, Marsh Stitchwort, Milkmaid, Mud Sedge, PF; Panicled Sedge, M; Purple Moor Grass, PF; Ragged Robin, Rannoch Rush, Round Leaved Sundew, Scots Pine, P; Tree Birch, M, P; Tufted Loostrife, Water Forgetmenot, Water Horsetail, M; Yellow Water Lily, M.

MOSSES AND LIVERWORTS

Calliergonella cuspidata, Calliergon giganteum, M; *Aulacomnium palustre*, M; *Bryum pseudotriquetrum*, M; *Drepanocladus vernicosus, Helodium blandowii*, M; *Marchantia polymorpha, Messia longiseta, Mnium rugicum, Mnium seligeri, Paludella squarrosa*, M; *Sphagnum subnitens*, M; *Sphagnum flexuosum, Sphagnum teres*, M.

WEED SPECIES PROBABLY BROUGHT IN BY WADING BIRDS

Barley, PF; Burdock, Cocksfoot, PF; Corncockle, Long Leaved Plantain, PF; Oak Seedling, PG; Sainfoin, Sharp Dock, PG; and Wood Avens.

Water Iris or Yellow Flag

Water Lily

Water Vole with Bog Bean

Common Frog

Ramshorn Snail

PLATE 20

The lists give a floristic analysis based on twelve such sites together with the sub-fossil evidence from the Book of the Motorway. The similarities are so striking that I feel further comparisons are unnecessary.

One further observation is however of great interest. I have returned to these shrinking lakes on three occasions after an absence of some five years in each case, and what has amazed me most is the rapidity of change. Over the period of fifteen years seven of them have virtually disappeared. The floating mats are no more; they have become stabilised, their surfaces colonised by a whole range of bog mosses, luxuriant growths of Heather and, in places, young trees of Scots Pine. Knowing the signs to look for it is great fun to wander through the local forest and with nothing more sophisticated than a long arm, reach down through the litter of pine needles to bring up sub-fossil evidence of *Paludella* and its close associates of the past.

Another reason I like to return to these sites is that if you are careful and use the right strategy it is possible to see the Grey Crane close at hand, or rather close at foot. In the summer the brown moss and peat stained water absorbs heat very efficiently, so much so that the surface waters are warm enough to take a bath in. So it is possible, choosing the right spot, to let yourself depress the mat and sink almost out of sight. From such a disadvantage point I have watched the large grey birds elegantly pick their way, balancing on the mat of floating vegetation, stopping to sun themselves and preening their long flight feathers. The edges of the water along which this and many other birds seem to like to take their constitutionals are often marked out by the presence of certain plants which seem very much out of place. This is a favourite habitat for the Great Fen Ragwort and the Bur-Marigolds, but they should be there: however, immixed among these denizens of the water's edge are Cereals, Docks and Plantains and on one occasion I saw Sainfoin in full flower in such a place. The most likely explanation for the tenuous presence of these aliens is that they have been seeded in by preening birds, or have indeed passed through their guts to be deposited ready fertilised in an alien environment.

So it was that many of the plants were carried north by the migrants seeking out new nesting grounds as the Great Spring made it possible. I wonder, did the cry of *Grus grus,* the once Common Crane, ring out over the Dale as their V shaped flights, each bird it's neck and legs outstretched, came in from overwintering way south in the warm Mediterranean? If so, perhaps where now the traffic roars along the motorway it was once possible to witness their mating dance, which is one of the most spectacular on record and perhaps it was they who brought the Sainfoin to the area where even today it finds its northern limit within sight of the same motorway some ten kilometres to the north.

These are but flights of fancy; we have no proof, but we do have undeniable evidence that the plant communities did exist right here upon the spot now covered with two broad carriageways of concrete. An arm of a lake created by glacial action was about to disappear in the advancing fringe of forest, for at that time the shores of the lake system were already clothed with open woodland of Birch and Pine with Willow and Hazel blown in for good measure. The pollens in the peat layers undoubtedly state that the first chapters of this particular history book were laid down when the climatic conditions were right and the process of succession far enough advanced for trees to grow at least within the lowlands of the Dale. So what were the earlier phases really like?

At about the same time that the construction of the motorway opened up this particular piece of history, excavations some forty kilometres to the west on a terrace overlooking the river, close to a point where it stops flowing within the confines of the Upper Dale and begins its more meandering course across the lowland plain, revealed another living history book. The deposit was found in undulating terrain of kame and kettle origin, the depression in question being some eight metres deep, carved into the glacial till, which had been made water-tight with a layer of stiff grey clay, then filled successively by marl, lake muds and silts, finally being capped with a thin layer of fibrous peat and sealed by mineral soil washed down from the surrounding slopes.

The basin and the story it tells are very different from the motorway site. Its

Hazel Bushes   *Dormice*

PLATE 21

sides were very steep, a common feature of such kettle holes, making the development of rooted aquatic vegetation difficult. Likewise its position, close to the edge of the river and within the grip of the regional watertable, meant that whatever communities did develop were subject to a continuous flow of ground water with at least a modicum of dissolved oxygen in it, thus preventing the formation of peat. Nevertheless the evidence is there, firm in the clays and silts; the print of the pollen is clear for those who have the ability to read it.

The ice had gone, the kettle was brim full of water and the first plants which left their mark within the aquatic scene were Pondweeds (*Potamogeton*) and Stoneworts. The latter being algae left no pollen to record their presence but in its place are neat oogonia, each one not unlike an old fashioned straw beehive or skip. Preserved in the clay marl interface is abundant pollen of Birch, Hazel, and Willow, with a little Pine and Juniper together with that of many plants of lesser stature.

It is the interpretation of such pollen spectra that provides the palynologist with his or her main problems and at the same time spices the science with the excitement of a treasure hunt.

Hazel relies on wind to carry its pollen to the female flowers, a somewhat haphazard process which is in part compensated for by the release of abundant pollen from its conspicuous catkins. Thus it might be expected that Hazel pollen would be present in great abundance, or if present only in small amounts could all have been brought in from afar. Those same winds that brought the first seeds and spores into the Dales could well have brought tree pollen up from way down south — long distance transport. The same is true for Pine, each pollen grain of which has the added advantage of two air filled sacs which buoy them on their way.

To interpret the pollen code it is thus necessary to count the grains of each and every type, expressing their abundance as a proportion of the whole or some readily understandable part of it. It is the painstaking though absorbing task of many hours. After preparation of the sample to highlight the presence of the pollen the scientist sits slowly moving the sample back and forth across the stage of

a well adjusted compound microscope counting each grain and scoring its presence on a sheet of paper: *Corylus* 1111, *Salix* 1, *Corylus* 11, *Betula* 111, *Plantago*
media/major (it is impossible to tell the difference) 1, and so it goes, hour after
hour building up a picture of the past. The excitement of the whole process is to
find a new identifiable grain or perhaps one you did not expect or haven't seen
before: the reward, the completed pollen diagram and notes for future reference.
In a sense a pollen diagram is never finished, for as new knowledge and techniques
become available that unknown grain may become the key to future understanding so samples, slides and notes are filed away within the ever growing archives.

The picture that emerged from the Mid Dale site was of a tundrascape, the
vegetation of which included Dwarf Birch, Ribwort and either or both the Great
and Hoary Plantain, Salad Burnet, Heather, Rock Roses, Nettles, Meadow-
sweets, Mugworts, Crow Berries, Meadow Rues, Willowherbs, Docks and
Thistle (the plurals indicating that identification was only to the generic level),
and members of the Bedstraw, Buttercup, Campion, Daisy, Goosefoot, Mint,
Rose and Umbel families, not to mention the grass and sedge pollen which
dominated the record at the basal level. The presence of these plants (the majority
of which grow in more open ground and often on bare soil), the dominance of
sedge and grass and the fact that the tree pollen, including Hazel, made up less
than 50% of the total count indicates that we are looking at the earliest days of the
Great Spring, Zone I of the Late Glacial.

What then were these open tundras really like and what animals found
sustenance from their summer bounty?

Without doubt the most typical herbivore of the tundra is *Ovibos moschatus*,
the Musk Ox. I will never forget my first encounter with a herd. I had been
making my silent way across a broad valley almost in the centre of Bathurst Island
(Northern Canada) making lists of the tundra plants which grow, in places, in
great profusion. It was a warm day except for a brisk wind into the teeth of which I
was, fortunately, walking. Earlier in the day I had come across a graveyard of
Musk Oxen, the remains of six animals littered across the centre of a large ice

AD 1000
BC 1000
BC 11,000

Dwarf Birch
Birch
Scots Pine
Elm
Oak
Alder
Willow
Juniper
Hazel

Grasses
Sedges
Cross-leaved Heath

AD 1000
BC 1000
BC 11,000

THE POLLEN RECORD
THROUGH 12,000 YEARS.
(not to scale)

PLATE 22

wedge polygon. The vertebrae were enough to give their identity away, some with the long spine-like apophyses which had in life anchored the large neck muscles so necessary for holding up the massive head complete with handle-bar of horns. It was of great interest to see that within the theatre of death and decay all the plants were doing much better than those growing round about, fertilised by the nitrogen, potassium and phosphorous still being leached from the skeletons. Two plants appeared to have gained special benefit. The first was *Carex bigelowii* which grew in great luxuriance in the corner of the site where water draining from the centre of the polygon would collect in the depression of the ice wedge itself. The other was a moss which grew upon the base of the horns which protruded from a covering of other commoner mosses. The only other place I had seen the moss *Tetraplodon wormskjoldii* was high in the Upper Dale growing on peat close to the remains of a dead sheep where it had been first found, not I hasten to add by the same sheep, but in the vicinity way back in 1901. Its tall soft tufts and ripe capsules each with an open trumpet bell-like end made me feel not far from home.

It has been a perfect day and suddenly there I was not 20 metres away downwind face to face with a real live herd drawn up in their typical defiant ring of confidence. The sheer bulk of the Musk Ox, plus its enormous coat of shaggy hair, is not well suited to rapid movement and certainly a lone animal might succumb to a pack of hungry Wolves. However a semi-circular phalanx of these creatures, pregnant females and young to the rear, adult males magnificent in stature and in stance, would be too much for anything on four legs, however many tails were wagging on their side. *Ovibos moschatus* must have been king of the tundra domain until Man came upon the scene. It was a sight never to be forgotten and one that made me realise that the graveyard in the morning had been perpetrated by Man, Eskimos out hunting for their food.

We know that Musk Oxen roamed the cold landscape of post glacial Europe; their unmistakable bones dated to this period may be seen in many a museum. Their possible appearance on the land so soon after the disappearance of the ice is easy to explain, for slow moving as these creatures are it is their nature to migrate back and forth, north in summer, south in winter, thus reaping the benefit of broad pastures. Such annual migrations would take them gradually further north each year and up from the plains into the valleys and dales as the ice retreated to the highest corries finally to evaporate into the warming air.

Their disappearance from the face of Europe must have been at the hand of something which could break the defiant shaggy ring with weapons and strategies made of sterner stuff than hair, hide and inherited behaviour. Yes, Man was among the new pioneers pressing north in the wake of melting ice and into the bounty of new spring. There were, however, many more great changes to be wrought upon the ice worked land before Man himself began to mould those same landscapes to suit his own special needs.

Why then, if the changes were coming on apace, is this brief period of our landscape's history (nowhere did it last more than two thousand years) still classified by the purists as Late Glacial, that is as part of the Great Spring rather than the beginning of the Great Spring? Could it be pure pessimism within the ranks of palynology, a syndrome brought about by the seemingly endless days of counting, before the spring of realisation when the completed pollen diagram reveals the whole history of that place? No, it is based on the evidence gleaned from the early chapters of many living history books across the face of Europe, evidence which indicates that this section of the world had not completely shrugged off the shroud of winter.

The pollen in the profile clearly shows that over the period during which the first marl was being laid down in the little lake the climate was improving: the kettle and its surrounding kames were warming up. As it did, Dwarf Birch which had dominated the scene around the basin gave way to Juniper in abundance and some Tree Birch (possibly the Odoriferous Birch which today has a northern montane distribution and is impossible to distinguish by its pollen from the others). The trees and other plants and animals of the forest fringe were marching north, bringing with it a profusion of ferns whose spores almost mirror the rise of

TAWNY OWL

PLATE 23

Juniper pollen and as they do so the pollen grains of many of the plants of open ground themselves begin to show a marked decline.

Two other woody plants which may have grown to the stature of tall trees or been no more than mature bushes, or even in the latter case dwarf species, are Hazel and Willow. Of the Hazel there is little doubt of its exact identity, for *Corylus avellana* is the only species which is commonly found in north western climes today although it does have fourteen other close relatives in the north temperate zone.

The Willow pollen presents a more difficult problem for it may have come from a number of specific sources. The Dwarf Willows which rarely grow more than ten centimetres tall are true denizens of the open tundra. The Tea Leaved Willow grows to around four metres in height in sheltered spots beside rivers and streams, where it may hybridise with the Dark Leaved Willow which grows luxuriantly in slightly drier places. There are of course many more and all produce similar pollen. If Willows of whatever stature were there so too must have been the insects which bring about pollination. The individual flowers on the Willow catkins are each supplied with nectaries which produce the sugary solution that attracts the insects which inadvertently pollinate the plants.

It is one of the great delights of the High Arctic to be able to see the annual seasons mirrored in the growth of these prolific plants. Their often colourful catkins develop, each one turgid with promise, either before, during, or after the first flush of new green leaves, depending on the species. Each catkin becomes the centre of buzzing activity as insects crowd in to gorge on nectar and fly away weighed down with pollen, some of which will brush off at the next landing which may be on a mature female flower. All Willows smack of male chauvinism because each male flower is furnished with two nectaries while each female often has but one. Next time you go for a walk in the spring take a magnifying glass and look into this world of miniature endeavour. The superiority of the male is however short lived: once the pollen has been released the male flowers die. Each of the female flowers in which pollination has been successful swells to produce a ripe capsule

containing many seeds, each of which is supplied with a white lint. When the
capsules burst open the whole scene may turn white as snow. The insects' work is
well done and the wind takes over, dispersing the seeds, carrying them on silken
parachutes to seek out new opportunities wherever they may land.

The green leaves have soon done their annual job, fixing the energy for
growth, for reproduction and enough to store away to tide the perennial tissues of
these tiny trees over the darkness of the next winter. The leaves thus lose their
chlorophyll, and the other pigments, yellow, orange, red, begin to show through,
washing the tundra in its autumn tints. Finally the biting winds of winter blow the
dead leaves away to form a rich blanket over the surface of the developing soil. The
cycle is over, a cycle which in the High Arctic may last as little as six short weeks.

The summer days of the Great Spring could have been as short, for late and
early falls of snow would produce a blanket as dark as any arctic night. However
the evidence, such as it is, would point away from this conclusion. The presence of
a large though diminishing ice cap would be sufficient to generate and maintain a
fairly constant anticyclone, especially during the summer months when the
temperature gradient from the ice cap to the surrounding ice free area would be
greatest. Heavy cold air accumulates over the ice cap and it would be unlikely that
depressions coming in from the Atlantic would make inroads into such a high
pressure system; the anticyclonic winds would thus be cold and dry and summer
rainfall minimal. At this time much of the ice free southern North Sea was high
and dry and so easterly winds would also bring little water with them. Cloudless
skies and a cold east wind would thus be the dominant features of the Late Glacial,
arguing for a long snow free summer period each year.

Whatever the length of the growing season there is no doubt that in the first
days of the Great Spring, Willows played their part in painting the landscapes with
the varied colours of the annual seasons. Their pollen is there, building up to a
peak as that of Dwarf Birch started to diminish. The Willow build up is followed
by Hazel, Juniper and finally Tree Birch.

These pollen peaks represent waves of new endeavour washing across the

Otters below the Waterfall

PLATE 24

landscape, each successive one gradually pushing further north, the glaciers melting back and up, the conditions improving in their wake and the vegetation following hot, well at least warmer, on their heels.

The bottom metre of the deposit, though just clay and mud to most people, is in fact a time machine allowing the succession of events over a thousand years of important change to be read with accuracy. The problem, however, comes in the interpretation of what was cause and what effect. Was the ordered succession of changes, recorded by the pollen caused by gradual climatic improvement or was it the gradual process of vegetational succession?

Well, what are the facts? It takes an awful lot of heat to melt a glacier, and so it seems reasonable to assume the climate must have been warming up for a considerable period before the terrain was free from its ice burden. Once uncovered, conditions would be right for the growth of a whole cross section of arctic plants including the hardier species of trees. The presence of ground ice, cold, dry north easterly winds and frost action, could have prolonged the arctic effect, maintaining the conditions of the open tundra. However, once the last vestiges of ground ice were gone colonisation by trees could be rapid.

It is not, however, that simple; the mineral material, be it bare rock or finest silt, must be prepared by the processes of pedogenesis and succession before true forest could develop. The rate of such change would depend, at least in part, on the nature of the parent material, bare bed rock taking longer than a well worked till but nevertheless it would take time. Comparison with modern secondary succession on cleared sites is of little relevance as much of the 'magic' of soil formation will have already taken place and so such processes should be much more rapid.

It must be made quite clear at this juncture that the evidence for these ordered vegetational changes does not come from this one location. Pollen studies from a whole range of sites across Europe allow the identification of similar pollen zones developed in a similar order. This in itself would indicate that the vegetational changes were in response to changes in the overall climate, a warming phase being designated Zone I and the following Birch dominated cool temperate period being called Zone II. Natural processes of succession and soil formation could account for the changes observed, but it is likely that they would be out of step time wise; for example those described above for the Mazurian Lake District are still going on today. What was needed was a method of obtaining real dates for the pollen events detailed in the different deposits. It is fortunate that the exciting and exacting technique of radio carbon dating came to the aid of palynology, so advancing the science and our understanding of past events.

If you were able to see and identify the components of the air you breathe you would find that three in every thousand were molecules of carbon dioxide. The air you breathed out would be considerably enriched with this gas for like all living things that use chemical energy you produce carbon dioxide as a waste product. The overall process is called respiration. All green plants rely on carbon dioxide in the air or dissolved in water as a raw material for photosynthesis, the main product of which is sugar. Glucose contains six atoms of carbon, and cellulose, which makes up the bulk of most plants, is itself made of many thousands of molecules of glucose. Add to this the fact that carbohydrates, fats and proteins, the major constituents of both our diets and our bodies, however ample, all contain carbon and the key importance of the carbon dioxide of the atmosphere becomes immediately obvious.

Although we cannot see the molecules of carbon dioxide, atom age technology tells us that one in every thousand billion of them contains radioactive carbon. What is more, this infinitessimal proportion remains more or less constant throughout time, regenerated by cosmic bombardment. As with all radioactive substances, once formed, carbon fourteen, as it is called, decays away with time, losing the radioactivity and hence its detectibility over a period of many thousands of years. The same is true of the carbon fourteen once incorporated into plant and animal material and that includes the likes of you and me: the processes of radioactive decay go on losing half their strength every 5,500 years. So it is that measurement of the residual radioactivity of sub-fossil material should provide a

MUSK OXEN

PLATE 25

method of dating the past, and taking all things into consideration it does so with an accuracy of a century, more or less, depending on its age and other factors.

Using such atomic age techniques it is now possible to state with quasi-confidence that Zone II lasted a little over 1,200 years, coming to an end about 8,800 years before the birth of Christ. It is, however, not so easy to put a date on the start of Zone I as the deposits are often very indistinct, having been mixed by frost action; also the exact date of the final melting of ice and subsequent growth of plants containing carbon fourteen must have differed across the north-south transect. Dates obtained so far thus vary between thirteen and fifteen thousand years ago.

A picture emerges of an improvement in climate which brought to an end the reign of white death, the actual disappearance of surface ice being the most important event taking place at different times in different areas, depending on their latitude and altitude. The continuing processes of climatic improvement, succession and migration brought the main events of vegetational development into step.

The first zone was thus one of change, the landscapes slowly dominated by tundra vegetation, the second zone a period of much greater stability which lasted for at least twelve hundred years. During this time the climate was cool and temperate and the vegetation best described as park tundra which consisted of patches of open forest Birch, with an understorey of Juniper and Hazel and perhaps Pine in places, interspersed with more open tundra communities, especially on exposed bare rock where succession would have moved more slowly. Meadows rich in sedges and grasses predominated in the flatter, wetter areas and the fact that their pollen dominates the spectra throughout Zone II shows that they were there in great abundance. Ample grazing thus existed for large herds of Musk Ox which may well have been there and for Elk which we know for certain were present at that time. Proof positive came from a brick pit not far from the motorway site where excavations in 1916 revealed a skeleton of *Alces alces*. It had been preserved where it had evidently died in amongst its food plants. These

included, in order of submergence, Bogbean, Bottle and Pond Sedge, Pond
Horsetail, Shore Weed, Pond Weeds and Water Millfoils. Dating of its death bed
by pollen analysis placed it firmly in the Late Glacial and by carbon fourteen
techniques 8,851±630 years before the birth of Christ.

One of the best places I know to see the Elk doing its own contemporary thing
in the wild is in that enormous stretch of marsh and peatland which spans the
borders of Poland and Russia. They are a westward extension of the Pripet
Marshes which is, at least for peatnicks like me, a mythical promised land where
large tracts of peatland still exist in their undrained natural state.

There is one magical walk through that country which I have made on a
number of occasions where the scenery must be very like that seen by our Elk
between his last mouthfuls — open marshland with reedbeds separated by
sedge meadows of every shape and size. Here and there these wet communities
give place to stands of Birch and on mineral islands protected from all but the
highest flood levels Pine provided welcome shade and a place to dry your feet.
When I made my first pilgrimage into this part of the present past there was a
raised trackway to guide the visitor over the wettest patches and I had been warned
that a member of the local fauna regarded it as their right of lay. The animal in
question was the European Adder and they were there indeed in strength sunning
themselves on the raised trackway. They were there also in length for their
numbers included the largest I have ever seen, carefully measured at 92 cms.

Many accounts say that Adders are timid creatures and will always glide away
at your approach. I must say that I have not found this to be the case in wetland
areas where they always seem loathe to leave their high and drier castles and so sit
tight. I hasten to add that they are never aggressive; all they do is sit and hope that
the intruder goes his own well chosen way.

The last time I visited that place much of the track had fallen into disrepair,
the wood was rotten with age and had part disappeared in the growing peat.
The Adders were however still in residence, their presence concentrated on those
lengths which still remained. Perhaps the local huntsmen had been deterred by

Viviparous Lizards with Cross Leaved Heath and Lady's Mantle

PLATE 26

the additional hazards of the track and like me been forced to take the wetter, cooler way and that is why upkeep of the track has lapsed.

I must in passing say that it is unlikely that reptiles of any sort reached the Dale in Late Glacial times, for being cold blooded, and hence completely dependent upon external temperature for their activity, they would have been likely to have been amongst the later of the post glacial immigrants.

Another animal which I always associate with this particular scene is *Tabanus sudeticus*, the Elk Fly. They are in fact the largest of European Horse Flies and when they alight they leave their own distinctive mark upon the place beneath, a neat, round, red puncture in the middle of a rapidly swelling lump. Elks must gain at least some protection from their thick covering of hair and from the fact that they can stand partly submerged in water while they are grazing. I must admit that I have followed their example on several occasions to avoid being driven to anaemic distraction by the ravening hordes of these and other blood sucking insects. It is however well worth the sacrifice if it allows you to see their normal prey at close quarters.

Elk are the largest of all the Deer and a magnificent male, which may top 800 kgs in weight, is adorned with a spread of hand-like antlers with many bluntish points or tines. The females are smaller and though they lack antlers they have the same soft muzzles which protrude over the lower jaw, giving them a quizzical look and making them noisy eaters. I remember sitting entranced, quite still for almost two hours as a family group browsed around me, close enough for me to hear the gentle smacking of their muzzled lips and feel more than my fair share of their flies. Their main diet appeared to be the young shoots of trees which were taken in preference to the abundance of sedges through which they trudged on their broad spreading feet.

Amongst their potential food were four rather special plants all of which may have been present in the Late Glacial landscapes of the Dale. First and foremost, Europe's most flamboyant Orchid, the Lady's Slipper, which certainly grew within the catchment of the Dale until the collectors of Victorian Britain transferred its declining populations onto herbarium sheets. Likewise Jacob's Ladder was present in the Dale until recent times, being last seen in the late 1800s. *Carex chordorhizza* and *Betula humilis* are the other two species in question. The former is known in Britain only from a few sites in the far north west of Scotland where it grows around the edge of lakes in lime rich country. The latter is unrecorded from Britain. Both are however very typical plants of the border between open sedge meadows and islands of Tree Birch and are often found associated with *Helodium blandowii,* another moss which has a similar distribution to *Paludella squarrosa* and indeed was found with that moss in the motorway site. The facts continue to fit.

Whatever the range of associates, both plant and animal, the period of stability with its cool temperate climate and park tundra which bears the name Allerod Interstadial lasted for no more than twelve hundred years. Allerod is the name of a town in Denmark where the succession of deposits that divide the Late Glacial were first recognised in a pit in a tile works. Interstadial signifies a short warm spell between two colder periods and that is exactly what took place right across Europe. Around 8,800 bc the climate once again became 'real cool', ice sheets re-advanced, glaciers re-developed in the highest corners of the Upper Dale and frost action was rife once again far to the south. To prove it, around the Mid Dale site the dominance of Tree Birch fell as Juniper and Dwarf Birch vied with the larger trees for dominance once again and Pine came in force for the first time. As the forest opened out and trees more able to stand the rigours of the cold moved back and 'flowered' in great abundance the lesser plants, which had typified Zone I with their presence, took on a new lease of life, some after an absence of more than one thousand years.

This cold snap, for that's all it was, dwindled away almost as rapidly as it had begun and Tree Birches once more grew and in their shade, lush stands of fern which may well have included the lime loving Marsh Fern, for the marl laid down at the time was much enriched with calcium, perhaps brought down by soliflux-ion. The continued absence of mollusca of any sort at this period perhaps points to

Elk in Tundra landscape.

PLATE 27

their renowned snail's pace or perhaps just to lack of data. Who knows but that a little kettle hole, set close against the mouth of the Upper Dale recorded this and the fact that the Post Glacial had come to an end, for there are the unmistakable traces of Oak pollen and Elm was soon to follow, indicators of warmer things to come.

So Zone III of the Late Glacial drew to its close, an event dated across much of lowland Britain at around 8,300 bc.

But what of the Upper Dale in which our story is unfolding? The Mid Dale site lies almost on the 200 metre contour while the Upper Dale will for the purposes of our story start at the foot of a great waterfall which is at 260 metres above sea level. It seems safe to conclude that during the Allerod Interstadial the glacier that occupied the Upper Dale melted away and life flooded in to take its place, but where and what and how much later?

If comparisons with the Southern Hemisphere are valid and there is no reason why they should not be, there is ample proof that the demise of a valley glacier set in a temperate climate may be very rapid. The Franz Joseph glacier in South Island, New Zealand has retreated more than 2,000 metres since accurate measurements were begun in 1894. However during this short time there have been three periods of significant advance lasting between two and fourteen years. In fact between 1965 and 1967 the snout of the glacier re-advanced a total distance of 300 metres, fed, it would appear from the records, by excess snowfall during those winters. The picture of such rapid regrowth is complicated by the fact that the glacier can itself slide forward as it melts and indeed such short term advances brought about by crevassing higher up may well signify collapse rather than regrowth.

One factor which speeds the final collapse of any glacier is the dust burden which accumulates upon its surface. As melting continues the dust and other debris which was spread throughout its volume is gradually concentrated at the top, turning the white reflective ice into a dark heat absorbing surface.

I have stood high on the Franz Joseph against the backdrops of Mount Cook

and other monumental nunataks, overawed by the river of frozen water filling the
valley with its ever moving presence. The pure white of the new firn, high on its
generative centre, contrasts with the dark, dirty mixture of rotting ice and debris
which marks its present line of retreat low down in the valley. The noise of ice,
rock and wind tell of the massive process of erosion.

From such a vantage point it is possible to see and understand the problems of
interpreting the pollen spectra. In favoured places sheltered from the wind and
filled with a diversity of debris, scrub woodland rich in species is able to develop
almost as soon as the ice is gone and magnificent forest clothes the flanks of the
lower valley, the bottom of which is still part filled with ice.

Gradual climatic improvement and/or the gradual process of succession?
That was the question. Perhaps it is of no more than academic consequence and all
one needs to say is that the final melting of the ice from the Upper Dale took place
later than lower in the valley. It is of interest to note that a rate of melt similar to
that recorded in New Zealand would have cleared the valley above the Mid Dale
site in as little as 2,000 years and from the motorway and Elk deposits in around
4,000 years.

To date the earliest definitive proof of vegetation in the Upper Dale has been
put at 8,120± 190 years bc. It comes from a peat deposit close beside the river some
15 kilometres from its source and around 60 kilometres (dare I say 3,000 years of
melt) from the lowlands where carbon dating puts the start of Zone I at between
11,000 and 13,000 years bc. This may be no more or less than lucky coincidence
but on the evidence to hand, and that is all we have, it would seem possible, and if
the facts fit they should perhaps at least be worn until outmoded. Oh for more
information and more accurate dates!

What then were the first plants to enter the Upper Dale? In order of
abundance of their pollen in the peaty record they were Sedges, Grasses, Willow,
Juniper, Dwarf Birch, Pine, Tree Birch, Hazel, and the rest all lumped together,
Elm, Oak, Jacob's Ladder, Starry Saxifrage, Mugwort, Meadowsweet, Dovedale
Moss and Moonwort. A mere fifty centimetres of peat yielded this precious

Barn Owl

Ivy

Crane Fly

PLATE 28

information and dating of its upper layers revealed that it took no more than fifty years to form. The ice was gone, the potential was there and the living history recorded the facts for posterity; but how may they be interpreted?

The abundance of the first five plants in the list would place it fair and square into Zone I and then the sequence would be clear, the first flush of the Great Spring arriving in the Upper Dale some 3,000 years after it had painted the lowlands green. The presence of Oak and Elm, even if attributed to long distance transport, appear to cloud the picture indicating much warmer times; late Zone III or even later and hence the vegetational development was in step time-wise throughout at least this part of Northern Britain, upland and lowland.

The logical explanation thus becomes to view the pollen zones as being of no more than local significance; what happened upon a certain place at a certain time determined by the conditions prevailing. Although they have been of immense help in the past the idea of pollen zones as having wider meaning both in space and time is best replaced by the concept of an 'assemblage zone', a dated sequence which spells out what has happened in that locale alone.

So we may conclude that the ice did not disappear from the Upper Dale until around 8,000 years bc and the first vegetation was a mixture of everything then on local offer; a mixture of plants of 'assemblage zones' I, II, and III, which had developed in ordered sequence lower down the Dale over a period of some 3,000 years, during which time the Upper Dale was still part filled with its rapidly melting valley glacier.

I would like to bet that no organic deposits containing records of vegetation from earlier late glacial times will be found within the Upper Dale. I hasten to add however that it would be great to have more information from more sites, especially with more accurate dates so that the palynologists could complete writing the history of this very special place. But why is it so special?

It was Sir Harry Godwin who pioneered pollen work in Britain and first cracked the zonal code, allowing the living history books to be read with ordered understanding. He first pointed out that the special plants of the Dale could be

relics of that first pulse of spring almost 12,000 years ago, relics of the past held as
it were in 'cold storage' by the upland climate, safe in an island castle set in a sea of
climatic change which would cover the rest of England with lush forests where the
light demanding plants of the Great Spring could not survive. A logical explanation but as we shall see there is ample evidence that those same forests invaded
their refuge and as the Great Spring turned into the Great Summer they filled the
Upper Dale with a blanket of deep green shade which meant death for the plants of
the open tundra. That was not all; changes even more deleterious to their survival
were to come and yet they struggled through, surviving all the onslaughts of
change right up until the present day. Theirs is a story of survival and it is their
story that makes the Upper Dale such a special place. So before we turn the pages
of those pollen books further and read on into the promise of summer I feel now is
the time to introduce the survivors, the *Dramatis Plantae* of my story.

Patterns of Maturity

Spring stands green with promise and summer terminates in a blaze of colours
which, though radiant in themselves, are fast muted by senescence. It is the
interface between the two which I find almost impossible to comprehend, for
spring drifts into summer, a continuum of living change with nothing to mark the
point when metamorphosis is complete. Likewise the drift of summer weather is
the most difficult to predict, for of all the seasons it appears more temperamental
than temperate in its progress. What is it then that marks this thing called
summer? That special busy heat bouncing back through buzzing insects from the
brown-green grass, the season in which even the best kept lawns of suburbia look
like some well worn carpet and gasp for water? Or is it that rumble of distant
thunder and sultry leaden skies announcing the onset of warm driving rain? No, it
is an amalgam of both and all the shades between, for even the worst summer
storm refracts into the arches of a double rainbow, distilling that sweet smell of
after rain renaissance. It is just that an enfleurage of warmth and wet leads to
crescendo and beyond to climax in an ever changing pattern of maturity.

Stoat through the Seasons

PLATE 29

The same was true of The Great Seasons, especially high in the Upper Dale where spring was late and with it came the trees of summer — Hazel, Pine, Oak, Elm, Lime, Ash and Alder, their own particular patterns of high life pulsing, changing, for the next seven thousand years. That special mix of warmth and wet brought the landscapes into climax, awash with the maturity of trees.

Every living thing re-enacts the seasons in its growth; at first the pace is slow, lagging as the sub-systems of macromolecules and cells mature until it is all-systems-grow. Then the pace quickens and stays that exponential way until accumulation of waste products, limitations of resource or is it just old age, slows down the pace, until senescence ends in death.

I like to hold an acorn, or come to that the seed of any tree, within my hand and think of all that it contains. Information, enough to make a forest giant out of nothing more sophisticated than the elements of the early philosophers: minerals of the EARTH, the gases of the AIR, the FIRE of the sun, quenched and made available by the special power of WATER.

Within the fruit the life of a new individual has begun; the slow growth of the embryo waits to be released through germination when the green of chlorophyll will take over from the stores of energy laid down within the seed, speeding growth on its way, seedling, sapling, maturity, and much more.

Each year the pulse of the seasons is recorded in the living trunk, the annual rings marking the change-over from the slowing growth of autumn, through the rest period of winter, to the urgency of spring and summer. These rings record the climate in great detail for they are made up of the cells which supply water to the annual crop of leaves which in return supply the energy for their growth. Any factor which alters this relationship of supply and demand will be taken down, clear evidence in the size of the conducting cells.

They also hold within their woody structure the evidence of carbon fourteen and any other chemicals which are incorporated from the environment into their working structure. It is through the study of such rings in trees of great antiquity and beams used in the construction of old buildings, that we are able to piece together a picture of the changing climates. Such study has also shown that the proportion of carbon fourteen present in the air is not as constant as was originally thought and hence it is that bc rather than BC is used to denote the inaccuracy of such dating.

As a tree grows it plays a more important role in the living landscape of which it is a part, providing food, nest space, safe sites, habitats in plenty for a multitude of life. Though it is but one individual of one species, in maturity it is much more than that; it is a living system, the complete study of which would take many experts much time. A tree, however large, is not immortal, for as its mass increases it requires more energy to keep that bulk alive and well, and eventually there will come a time when the balance between energy fixed by the leaves and energy used by the living mass swings from positive, through equilibrium and into the deep red. The growth of the forest giant then slows, stops, and soon it begins to rot away and die.

Sometime during its span of life it may produce one seed which will fall in some favoured spot where it will germinate and thus begin the same process of growth again, the new individual eventually replacing its parent in the economy of forest life. This point is rubbed home again and again each year as the vast majority of seeds and fruits are either eaten or are doomed to an etiolated death within the shade of the parent tree. It is no more than luck, a matter of pure chance which single offspring will survive to take its parent's place.

To me it is one of the most sobering of all thoughts that for any individual one successful offspring is sufficient, the job of life well done, for if there are more the surplus may become the stuff of which population explosions are made.

Such explosions were in order as the population of plants and animals spread north and many offspring of each hardy pioneer gained survival in the land rush days of the Great Spring and Summer. However, once each living landscape had come to climax, population regulation maintained the order of maturity and all save a mean of one per individual in each population perished in this harsh generation game.

Edible Dormouse with Emperor Moth

PLATE 30

If you want to witness this game in operation and sense the changing patterns of succession and of maturity, I suggest that you clear a section of your garden, remove everything from the plot, dig and turn the soil and then leave it to its own devices.

It is an experiment once started that you must leave, a legacy of interest for your own future generations, their own natural experiment in living. At first the plot will be dominated by herbaceous weeds, a youthful community the patterns of which will change as fast as youthful customs. These will then be replaced by shrubs and eventually by trees and thus each year will show an increase in the standing crop of vegetation and of humus in the soil. It is this increased storage of energy which mediates the changes of succession, altering the structure of the soil and the climate near the ground, creating new conditions, new environments in which other plants will grow and other animals can live. The patterns of the living community will continue to change until a state is reached in which there is no further net storage of energy; this state is called climax.

In climax all the energy fixed by the leaves in unit time will be used in the maintenance of the complete living whole. The rate of change will slow as the pattern of maturity becomes fixed. It is in the final stage that the living community differs from an individual. Senescence in an individual soon ends in death but a mature climax community may go on for many years maintained by a constant recycling of raw materials, reprocessing of toxic wastes and replacement of one generation with the next. Death and destruction only comes if the overall environment changes, conditions arise that bring recycling and regeneration to a halt or new plants or animals arrive upon the scene and find a role to play within a new pattern of maturity.

So it was to be throughout the Great and Glorious Summer; the more open communities of the Great Spring were gradually replaced by climax forest as wave after wave of trees advancing from the warm south arrived to colonise the Upper Dale.

I like to put myself in the position of the tundra plants and imagine the advance of the trees of summer like a series of invasions, with the strategy and long term success of each invading sylvan army dependent, at least in part, on their adaptability in the face of what was left by the last ruling class and the vagaries of climate during their campaign for dominance. Each one made ground, came into domination and then either retreated or became integrated into the local scene, playing a subordinate part in the new pattern of forest diversity.

To date eleven living history books scattered through the Upper Dale have given up their information concerning the Great Summer and although each story differs in its detail the overall plot remains much the same.

As always one of the problems lies in the interpretation of the data and the fact that the rate at which each chapter of a living history book is formed may differ, greatly dependent upon location, rainfall and indeed a multitude of local factors. As the peat forms, the annual rain of pollen is trapped to become a record of the past. If the rate at which peat is laid down is fast the pollen record will be spread thinly through the strata; conversely if the rate is slow the pollen will appear in great abundance and such variations must cloud the interpretation. What is required is either a peat deposit which has been laid down at a steady rate, and these are very few and far between, or a series of many radiocarbon dates, a process which is unfortunately expensive to perform. In absence of the exact facts of absolute (timed) pollen counts, the interpretation usually revolves around a comparison of the proportion of each type of grain in relation to the rest and thereby hang many tales of palynological woe. The fact is that expressed in this way an increase in the abundance of one type of pollen will produce what appears to be a decrease in the importance of the others, even though no real decrease may have taken place.

A new pollen grain may well come to dominate the spectra but that doesn't mean that the other plants were retreating from the scene. However there seems little reason to doubt that the first tree of summer to arrive in the Upper Dale was the Hazel, *Corylus avellana*.

First Trees of Summer

HETHER it is really fair to call *Corylus avellana* a tree, I am not quite sure: in strict botanical terms it is a phanerophyte because it bears its organs of perennation (winder buds to be exact) more than 70 cms above the surface of the ground. An arbitrary height set by some international code of nomenclature? No; a critical distance above which it is unlikely that those buds will be protected by a blanket of snow and so will remain exposed to the full rigours of the winter. The term phanerophyte however covers both trees and shrubs, thereby clouding not clarifying the issue. The books say that Hazel can grow up to 12 metres high and when it does it forms a handsome, slender tree. I must however admit that I have never seen one taller than 6 metres and that was a rather untidy, multistemmed bush. The explanation for this lies partly in an understanding of apical dominance and partly in the practice of coppicing.

The primary growth of any tree is localised in the tips of its shoots and roots, the actual area in which cell division takes place being called the apical meristem. As the plant gets older other lateral menstems come into action, slowly transforming the slender herbaceous stem into a woody twig, a lead branch and eventually into a tapering woody trunk. New apical meristems arise as new lateral branches are produced. Throughout the life of the tree the apical meristem at the tip of the topmost lead branch plays a very special role, controlling the growth of all the subordinate meristems and thus the growth and development of the whole plant. This is called apical dominance and if you want to see it in action then all you need to do is find a tree the apex of which has been damaged by lightning or some other natural catastrophe. If you can't find one then you could institute an experiment of

67

your own by cutting the top off a lead shoot, but please before you do, think carefully because it will have a drastic effect on all further growth. In the absence of a controlling tip one of the lateral branches will turn to grow upright and will take over as the new control centre. Some trees appear very susceptible to such treatment and will die if decapitated too many times, while others appear to thrive under exactly the same conditions. The latter is true of all the common tree bushes or bush trees of our hedgerows, and Hazel is one of these, the end result being a coppiced tree or a multi-stemmed bush. Long before Man started coppicing Hazel with a pruning hook however primitivie, herbivores were producing much the same effect with a well aimed nibble.

In compromising terms Hazel is a tall shrub with a tendency to grow on wetter soils except in limestone districts where it may be found in the driest of spots.

Much of the Upper Dale must have therefore offered ideal conditions for its colonisation, for the limestone, boulder clay and glacial tills would all have been rich in calcium, especially in the early days before the leaching power of the rain had taken its toll.

What then could a forest of Hazel have looked like? Well, the nearest thing I know of is in the Burren District of County Clare in Western Ireland, where a number of the rare plants of the Upper Dale are also found growing in abundance. These include Shrubby Cinquefoil, Spring Gentian, Mountain Avens and Mountain Everlasting and the presence especially of the alpine mountain flowers is made even more remarkable by the fact that they grow almost at sea level.

The Burren is a large block of limestone which, being exposed to the full force of the winds blowing across the Atlantic, is not the ideal environment for the growth of trees and indeed the only extensive woodlands are of Hazel and they are only found in the more sheltered areas.

From outside, the Hazel thickets look almost impenetrable; however, if you are willing to stoop and crawl on your hands and knees access is possible although it can be very slow. To date I have never had the good fortune to meet one of the

Bank Vole on Shrubby Cinquefoil

PLATE 31

Little People but I have often startled the bearded coppicers at their work in the guise of wild goats of which there are plenty. However the main coppicing is done by the salt laden winds which rapidly kill any apical meristem unwise enough to raise itself up above the surrounding shelter.

There is one spot in the Upper Dale close beside the river where there is a stand of Hazel which is large enough to accommodate even my ample proportions and again hands and knees are the only method of entry. The ground is carpetted with moss which itself supports the growth of a weird little fern called the Adder's Tongue, a plant which is not uncommon in the Burren woodlands. I like to creep in there in the early spring, sending up clouds of pollen from the long male catkins (no wonder it swamps the pollen diagrams) and again in the autumn when the cob nuts are ripe and ready for eating, or, if you are a Red Squirrel, to store away ready for winter.

It is in the autumn that the male catkins start their growth which continues only after a long winter rest. The female flower heads are minute compared to their chauvinistic counterparts, being no more than drab brown buds. However what they lack in size they certainly make up for in colour, when their crimson stigmas suddenly turn each one into an exotic miniature hanging garden, a real joy to behold through a times-ten lens.

Exactly when the Red Squirrels first arrived in the Upper Dale we shall probably never know, but in the knowledge that they thrive today way up north on the tree line, I feel there is no reason to doubt that they arrived soon after their favourite nut bushes. They are certainly still in residence and looking up to the stands of Mountain Birch which fringe this scrap of Hazel woodland it is difficult to make out their dreys from the abundant witches' brooms which adorn the Birch branches, especially when viewed against the sun. Squirrel dreys are gigantic balls of twigs and so are witches' brooms, the main difference being that in the latter the twigs radiate out from a living centre and owe their development to another breakdown of apical dominance rather than to the nesting drive of the Squirrel. They are the product of a parasitic bacterium which voids chemical waste into the living tissue of its host causing the disease whose outward manifestation is a witches' broom. The waste products of the bacteria are so like the normal control chemicals produced by the dominant meristem that the control is lost and the diseased shoot keeps branching and branching in an attempt to restore law and apical order. *Bacterium tumifaciens* is a parasite living off other forms of life. Red Squirrels are omnivores living off a variety of other food. Once Hazel had come to the Upper Dale in force these and many more living organisms could follow on and so the diversity of life increased.

The food of the Red Squirrel ranges from young shoots, seedlings, fungi, flowers, fruits and seeds, and though they pervade a 'cuddly' image they are not above eating birds' eggs and even defenceless nestlings. Contrary to popular belief, although they do lay down food stores for the winter, they do not hibernate for long periods. They may well retreat into the shelter of their dreys in very bad weather but as soon as it warms up they must be out and about feeding on whatever they can find or returning to search for their caches of food. So haphazard are these that some may be entirely overlooked and the buried fruits and seeds may sprout providing a mini forest of seedlings. The activity of the Squirrel may thus aid the dispersal of the plants.

So it was that Hazel made its way up into the Dale, first in the more sheltered spots but as the weather improved it soon covered much of the area with an open forest, smothering those tundra plants that cannot grow in shade. As the area of tundra diminished in size those grazers which are dependent on the tundra and in part on the special food, like the rich cover of lichens which the open vegetation provides, may well have suffered the same fate, especially in competition with more efficient woodland grazers which moved in with the trees.

The two large herbivores in question are the Musk Ox and the Reindeer, both of which thrive today only in the open park tundra and further north, although there is no getting away from the fact that this contemporary reduction of range may well be due in part to Man. We know that the Musk Ox was a member of the fauna of England before the last Great Winter and it is likely that it lived around

Red Squirrel

Ivy Leaved Toadflax

Black Slug

Longhorn Beetle

PLATE 32

the fringe of the ice sheet. It could well have been exterminated during this time by bands of hunters making summer excursions from the warmer south and may never have advanced north with the spring. However I am an optimist and as stated above I like to think it 'made it' to the Upper Dale and hope that one day we shall find proof of its gentle presence.

Reindeer certainly moved north with the melting ice and their antlers displayed in a number of museums prove the point. Their demise, especially if at the hand of Man is in some ways more difficult to understand because of all the large grazing animals they are the most eminently domesticatable. The Reindeer of Eurasia, and the Caribou of North America, are in all probability one and the same species of animal and as if to prove it they are the only members of the Deer family in which both the male and female have antlers. One explanation of this is that in the winter when the female must clear the snow to allow her young to feed on the lush carpet of lichens beneath, she must be equipped to guard her feeding 'crater' from other members of the herd on the look out for an easy meal.

The lichens of the tundra which include the misnamed Reindeer Moss (*Cetraria aculeata*) and complexes of Cladonia species which are often so abundant as to maintain large sections of the tundra landscape white even in the absence of snow, do not appear as the tastiest of morsels, at least to us human beings. However some of them can be made edible even to our fastidious palates by the right sort of cooking. To the Reindeer they are a staple part of their diets especially in the winter when the bulk of the other members of vegetation have died back, leaving nutrient-poor litter from which even the bulk of the important minerals have been withdrawn to be stored in their underground parts. Not so the tough lichens: they remain turgid (it would be wrong to say juicy but this is implied), almost unchanged in nutritive value throughout the winter, stored in not-so-deep freeze just beneath the snow.

Reindeer are nomadic, migrating long distances to and from their winter grazings and have been domesticated both in Lapland and Siberia by nomadic herdsmen at least since the fifth century. At first tame Reindeer were probably used as decoys when hunting their wild cousins but it was soon found that they thrived in captivity where apart from providing meat and hide they were good draught animals, being able to cover rough frozen ground with amazing speed and find their way in white-out conditions. Man thus learned how to 'parasitise' these gentle creatures, for a herd can be managed by a young child, providing the herd some measure of protection from its natural enemies and in turn reaping everything necessary for their nomadic life.

As the vastness of the contemporary tundra opened up, *Rangifer tarandus* must have migrated north and there seems little doubt that some would have found ample grazing on the tundras of the Upper Dale, but as their tundra home diminished in size with the advance of the trees they would have been gradually replaced by the herbivores of the forest. Nevertheless it is nice to think that perhaps the first Hazels were coppiced into bushy submission by the gentle grazers of the tundra.

One further legacy of the tundra was rich layers of humus produced by the abundant mosses, bog mosses and lichens that once dominated the scene. These had done their job, adding new chemicals to the armoury of the process of erosion, filling the cracks between frost shattered stones, forming the first skeletal soils. Bare broken rock inanimate but full of living promise, was ready to be colonised by any plants and animals which could stand the rigours of pioneer state and speed the process of colonisation, settlement and the formation of a living soil. Dead rock was now vibrant with life, rhizoids, rhizomes, roots, all reaching down to tap the rich supplies of water and mineral salts. The annual rain of leaves and twigs and the death of the underground parts themselves became food for a whole cross section of animal life: insects, worms, centipedes, spiders, mites, and snails to name a few, who came to the Upper Dale to reap rich harvests and to play their part in the formation of new soils and new societies of life. When each of the members of the busy burrowing society of the soil pass to rest their remains are broken down by hosts of bacteria, fungi and protozoa, the minerals they contain being put back into the cycle of the living soil. Just as important, a certain amount

PINE MARTEN

PLATE 33

of the rich humus is longer lived, being less tractable to decomposition, and so long as it remains it enters into quasi-union with the mineral particles, binding them to form into larger aggregations or crumbs. These being irregular in shape and outline lie together in a higgledy-piggledy of solids and spaces through which water can freely drain, and air can penetrate, giving the all-important roots, rhizoids, and creepy crawlies easy access while themselves stirring and speeding the process, the formation of a structured living soil. Water percolating down through such a structured profile will tend to leach away the available minerals down to the lower levels: deep penetration of roots and of worms will return them to the surface, boosting annual productivity, providing richer supplies of humus to play its many roles which include holding a certain amount of water and minerals within its colloidal grasp, where they are readily available to new plant growth.

The new cycles of life initiated by the invasion of Hazel, be it bush or tree, speeded the process as strong brown roots reached down and the soft hairy leaves added to the rich humus in the fullness of each autumn.

Put down the book: go out into your own environment, stoop and take up a handful of soil, feel its structure, sense the life within, and replacing it with care take counsel of the world's greatest and yet least appreciated resource; soil, the living coverlet of Mother Earth, the mediator between dead rock and all things animate that have taken up the challenge of living beyond the confines of the sea.

If the Red Squirrels did come to the Dale in Hazel times, burying their winter food in the developing forest soils, they would not have felt out of place in the Pine forests that were following hard on their heels. *Pinus sylvestris*, although locally known as Scots Pine, is widespread across Eurasia, penetrating south west as far as the high mountains of the Spanish peninsula. It is a magnificent tree which when full grown to 40 metres in height is truly majestic. Its long straight trunk and sparse crown of branches at the top make it instantly recognisable but it is the colour of its bark which is its prime distinction. Orange red? No, high up where the branches have their genesis there is an unmistakable flame pink. The whole collage, with the flaking of the bark providing a third dimension of colour and texture, changes with the moods of the sun. As it sets or rises the whole tree glows with its own ambience of majesty.

Today there are no Pines of any merit within the Upper Dale: their rightful place has long been usurped by Norway Spruce imported from the continent to be grown as a cash crop. Once it was a native of the British scene but that was before the last Great Winter cleared the forests, pushing their zoned presence to the south. Why Spruce did not make a come-back with the melting of the ice is still a mystery and again there are as many theories as thinkers in the field. Perhaps it was its intolerance of competition with members of the Heather family which are dominant features of the vegetation of both the tundra and the oceanic climate of the Atlantic coast.

In order to establish Spruce upon our fells today the land must be deep ploughed, and the seedlings planted on the upturned sods above the choking reach of Heather. Or perhaps Spruce was a slow mover so that its spread north was blocked by the waters of the English Channel and the North Sea, the landfall on the other side already colonised by trees more tolerant of youthful competition? Who knows? Suffice it to say that there is no shred of pollen evidence to suggest its natural presence. *Pinus sylvestris* made it before the migration routes were closed to become our Scots Pine. *Picea abies* missed out until much later when its introduction and establishment was aided by Man.

If you would have the immense pleasure of seeing a real forest of Scots Pine then a pilgrimage must be made to one of the areas in Scotland where its presence as King of the Royal Forests has not yet been usurped by the Pretender from Norway, aided from within by the Commissions of economic forestry. At Rannoch, Rothiemurchis and Loch Maree it is still possible to see natural Pine forest set against a backdrop of mountains the flanks of which bear open tundra-like vegetation.

My favourite is at Loch Maree in Wester Ross where it is possible to stand beneath its shade, suffused by the sweet smell of Juniper, with autumnal Lady's

*Wood Mouse with Wood Sorrell, One Flowered Wintergreen
(King Olaf's Candlestick) and Bluebell*

PLATE 34

Tresses, Lesser Twayblade, Single Flowered Wintergreen and Twin Flower about your feet.

Perhaps a word concerning the Latin naming of flowers is not out of order here, for the Twin Flower *Linnea borealis* bears the name of Carolus Linneus, Swedish naturalist and founder of the two-name system of Latin nomenclature. *Linnea*, in homage to Linneus; *borealis*, of the great northern forests or indeed of this period of time, for this part of the Great Summer in strict palynological parlance is called the Boreal Period. The Single Flowered Wintergreen rubs home the difficulty of using common names. In certain parts of Scandinavia, where it is a common plant, it is known by the much more descriptive name of King Olaf's Candlestick, and in other places by other equally evocative names. So to avoid confusion the world of strict botany used *Moneses uniflora*, and at the end of this book there is a concordance linking all the common names used in the text with their proper counterparts. Nevertheless I am sure you will agree that whatever you decide to call *Moneses uniflora* it could not be any more beautiful nor more perfect in its flowering detail.

Real Pine forest is the perfect place in which to sit and think of taxonomic problems, to transport yourself back into Boreal time or just to watch the Red Squirrels chase up the cracked bark, usurping the Crossbills from their meal of seeds held safe within the scales of the pine cones. Safe? Only until sharp teeth of the Squirrel cut their way through to the goodness within or the crossed bill of *Loxia curvirostra* prises open the wooden scales to feast on the bounty of winged seeds. Only when the weather is right do the cone scales open of their own accord, responding to the dryness of the atmosphere, releasing winged seeds which may be blown north, south, east or west to germinate wherever the conditions are right.

There is one main difference between the Pine forests of the Upper Dale and their modern Scottish counterparts. The former lacked Heather as an important part of their ground flora whereas in the latter Heather is often dominant. The natural Scottish Pine forests are open in character; the trees being uncrowded, light can penetrate with ease through the open canopy of needle leaves. However, over much of the area light cannot reach right down to the forest floor owing to the abundant growth of Heather which may grow to a height, or more meaningfully to a depth, of more than 100 cms. The life of lesser plants within the aegis of the stands of Heather thus come to depend on the cyclical growth of the dominant, which like all other living things follows a patterned path to maturity. Each Heather plant goes through four phases of growth — pioneer, building, mature and degenerating and in each it differs both in its ability to compete and in its susceptibility to competition. The dense stands of building and mature Heather choke out the more delicate members of the flora and may even inhibit the regeneration of the Pine itself, yet as the cycle of growth continues on its own relentless way the patterns of maturity change, opening up new opportunities for others to make their mark upon the scene. Plants like those already mentioned and many more, Cow Wheat, Chickweed, Wintergreen and the sinuous creeping Interrupted Club Moss and in the damper spots the tiny Coral Root Orchid sprouting from soft pinkwashed carpets of Girgenson's Bog Moss. If only one could sit across the years and contain the changing patterns in action-replay then you would see the pulse life, the interplay of species, the opportunities gained and lost, the process of regeneration providing continuity in time through change in space.

The pollen evidence tells us that all this must have happened in the Pine forest of the Upper Dale but that it happened in the absence of Heather. The presence of abundant lime in the still young soils is sufficient to account for the lack of Heather which is a lime hater and hence for much more open conditions on the forest floor. These conditions probably remained throughout the 3,000 years in which Pine reigned supreme in places, its own fortunes expanding and contracting with the changes of climate and the advance of other species.

The whole problem of forest regeneration is one of great interest and one that still vexes ecological thought. Conditions which must have been ideal for the germination and establishment of Pine within the open forests of Hazel around

Female Mallard

Water Mint

Banded Agrion Dragonfly

Frog Bit

PLATE 35

7,500 years bc may well have changed drastically when several hundred years later senescent trees were in need of replacement. Seedlings must often compete with their parents for water, minerals, and light, and it is often light, or rather lack of it, which limits regeneration.

Hazel produces a soft humus which in the presence of the right soil flora and fauna will be rapidly broken down. Pine in contrast produces a raw humus, its needle leaves full of resins and tannins which not only render them unpalatable to many of the soil animals but it also contain compounds which are bacteriostatic and hence slow down even the final steps of recycling decay. The consequent build up of raw acid humus must have benefitted the growth of certain plants and it is interesting to speculate on the reasons as to why Heather did not make an early appearance on the forest floor, as conditions changed and lime loving plants like Lily of the Valley must have been ousted from their former strongholds.

Thus it was that the micropattern of life on the forest floor not only was controlled but also helped control the macropattern of the forest stands themselves, the whole changing through time in an ordered though disorderly way. Like all patterns of whatever dimensions each consisted of at least three parts, the two or more units which make up the pattern and the boundaries between those units. An artist can, if he or she wishes, make a pattern as abrupt as may be desired, playing up the units and playing down the boundaries to the finest of fine lines drawn with the hardes pencil or painted with the best sable brush yet the subtleties of the boundary are still there, an intermix of carbon particles, or pigment and its vehicle, there to be appreciated by those who, rejoicing in the technology of art or science, can comprehend in microscopic dimension.

Likewise in nature no boundary is absolute. Even when viewed from the elevated position of *Homo sapiens* boundaries appear as a gradual intermixture of the members of the two adjacent societies of plants and animals which make up the units of the pattern. Each unit is in itself a blend of different populations, each made of different individuals and their different parts, a vital admixture of units and boundaries together providing opportunities for a fuller expression of life. It is in such boundary situations that the full dominance of the units can never be expressed and so they represent zones of instability, and opportunity within the stablising patterns of maturity. The broader the boundary the more gradual will be the transition from one sphere of dominance to the next and the more 'free space' for plants and animals which do not belong to the two adjacent communities to live, thrive and have their being. Patterns within patterns, units within units, boundaries within boundaries, instability within maturity; a continuum of environments changing through space and time, within which members of a ruling class long past may linger on or where new patterns of revolutionary change may have both their genesis and their exodus.

The boundaries between the Hazel scrub and the Pine forest must in the main have been one of retreat for the former and advance for the latter and yet the pollen record indicates that Pine did not have all its own way. The Red Squirrels could have survived in either, supplementing their diet of cob nuts with the bark of young Pine shoots or the larvae of the Tortrix Moth which bores into the soft reserves of food laid down in the young Pine buds. Thus they could have been both a positive and a negative force in the spread and survival of the Pine, one more factor in the increasing complexity of life in the Upper Dale. Within the Pine/Hazel border a number of the plants of open tundra times could have maintained their presence in the Upper Dale, coming into contact with new immigrants from warmer continental climes; plants like Hoary Rockrose, Horse-shoe Vetch and Rare Sedge. There too the vanguard of the new invaders may have made their presence felt, for it was around 6,800 years bc that Oak and Elm began their foray into the Upper Dale. The pollen evidence states quite clearly that they came together, the Oak in all probability moved by animals and birds, the Elm seeds floating on their brown membraneous wings which do not really work until the fruit and seed are completely ripe when, paper light, even gentle thermal winds can lift them up and carry them away.

At first glance the pollen evidence derived from some twenty living history books scattered throughout the Upper Dale is confusing to say the least, but

Fallow Deer Fawn

Wood Millet Grass

Bramble

Dusky Cranesbill

Small Tortoiseshell

PLATE 36

detailed study, especially when guided by the expert in the local field does reveal a clear picture of the forests of the Great Summer. (I have been more than fortunate while writing these chapters to have the constant advice of Dr Judith Turner of St Aidan's College in the University of Durham who has done the bulk of the palynological work in the Upper Dale.)

Hazel is always there, its fortunes fluctuating with the expansion and contraction of the forest types. The same with Birch, a mere background trickle of pollen grains, but always there ready to make a comeback if the opportunity arose, especially at the higher levels. Both these species have the ability to thrive on a range of soils from wet to dry and from lime rich (neutral to alkaline) to lime poor (acid), although it must be concluded that in conflict Birch would stand to win on the more acid soils and Hazel on the more alkaline. Then in came the Pine, overshadowing the lesser trees of the late spring; but it did not have things all its own way for the evidence clearly shows that it was confined at first to certain soils and initially made but little impression above the 500 metre contour. Some 1,000 years later it moved up in force, coming into patchy dominance up to, but never much beyond, the 700 metre mark. This 'hesitant' pattern of colonisation may well be related to its preference for the drier situations, its upward limit always being at least part controlled by the wetness of the climate which then, as it is today, must have been related to altitude. The final demise of the Pine came around 3,500 bc when the overall climate was getting wetter and the increased abundance of Alder tends to bear this supposition out. The fall in Pine and rise in Alder pollen indeed marks the end of the Boreal period and full discussion of it must be left until later. First we must consider the change-over from evergreen needle leaved coniferous forest to deciduous broad leaved forest.

During Pine's first thousand years of residence within the Upper Dale its advance was caught up and indeed overtaken by both Oak and Elm. In the knowledge that of all the trees so far mentioned Elm produces least pollen, it is fair to say that throughout this period and until the Elm itself declined into insignificance around 3,000 bc, both Oak and Elm were of equal importance in the Upper

Dale. It is also clear from the pollen record that Elm advanced more rapidly.

The picture of these Boreal forests is thus of Hazel and Birch occupying the least advantageous of sites, their abundance increasing with altitude; Pine occupying the best drained soils especially where limestone rocks outcropped through the wet covering of boulder clay; Elm and Oak making their way up through the more open boundary communities, at first in the more sheltered areas of the valley bottom but, eventually, to cover much of the flatter land. If any preference was shown it was by Elm which is, in competition, more tolerant of the wetter soils beside the rivers and indeed at the highest altitudes, for pollen evidence shows its undoubted presence on the highest fells of the Dale.

Across the Northern world there is even today a broad belt of similar mixed forest which girdles the earth, forming a ragged boundary between the true taiga further north and the broad leaved forests further south. The exact admixture and expression of needle and broad leaved trees within this zone depends on the whole complex of environmental factors — latitude, altitude, aspect, soil and land use history. Likewise the whole complex admixture of species population and vegetational units changes with the changing climate. Despite the complexities of this elaborate boundary situation it is possible to set some broad limits in environmental terms on the distribution of its main components. The trees of the broad leaved deciduous forest zone appear to find it difficult to thrive in situations where there are less than 120 days per year with an average temperature of above 10°C. Likewise the dominants of the coniferous taiga do not survive in areas with less than 30 days where the average temperature is above 10°C. Thus we may conclude that as the climate improved conditions became right first for the growth of Pine and later for the growth of Elm and Oak, but that at least for 2,000 years the Upper Dale remained a boundary zone in which all could find a place and in which no single species could come to overall dominance.

What is more, the rain of pollen tells us that the forests were open in character as there is throughout these particular pages of living history an abundance of pollen of herbs, grasses, sedges, and the spores of fern. It also seems safe to say

Wild Boar

Emperor Dragonfly

Pendulous Sedge

Northern Water Sedge.

PLATE 37

that all the forest types thinned out with altitude, the trees getting smaller, especially on the more exposed crests of the fells. Pine never made it to the highest altitude and yet there is firm evidence of Elm at even the highest elevation, its unmistakable pollen held safe in tiny pockets of peat which supported an abundance of Globe Flowers. These upland forests must have been pleasant places through which to walk and graze and many organisms found a niche within the Boreal forests, living off their bounty and governed by their patterns of maturity.

Wild Pig rooted for acorns and the corm of the well named Pig Nut, turning up fresh mounds of earth which could be colonised by Wood Millet, Holy Grass, the creeping presence of Yellow Pimpernel and the feathered heads of Wood Horsetail. I believe that Wild Pig would have been an early immigrant into the summer forests of the Upper Dale for they travel far at night in search of a little something to slake their legendary appetites. Apart from the older males who live a solitary existence they are social creatures living *en famille* for the majority of the year; gathering together in the autumn they may form bands of up to fifty individuals. If the conditions are right, open woodland with wet muddy wallows and plenty of food, the population of Wild Pig can build up very rapidly. The sow comes into season every three weeks and a successful mating produces a large litter in 112 days, throughout which time the mother, though not so agile as normally, is still capable of putting up a good fight. Survival of the piglets is, in the early days, simply a matter of obtaining a teat at feeding time, and with between eight and fourteen on offer this presents little problem except to the weakest. They are weaned at around twelve weeks of age when they change to the most omnivorous of diets. Wild Pigs will eat anything that comes their way: roots are a firm favourite including those of ferns which few, if any, other animals touch, fungi of all shapes and sizes, worms, insects, frogs, lizards, birds' eggs, the young of any other creature of the forest and when it comes to carrion they are not averse to cannibalism. Their numbers would have been held in check by lack of resources and by predators such as Wolf, Lynx, and Goshawk, although only the Wolf would have been able to tackle anything but the young.

I have recently had the fantastic pleasure of watching a Goshawk at work in Man made forests not all that far from the Upper Dale itself. Although about the size of a Sparrowhawk they look much bigger as they fly flat only amongst the boles of the trees, chasing birds on which they prey. The abundance of bones of both Rabbit and Hare around their main flight paths showed that they would tackle larger more terrestrial things.

Lynx must have added their feline presence to the mighty forests, burying faeces and urine within their chosen home range, yet marking its periphery both by sight and smell of those same excremental products. They are silent killers dropping from above to kill a small deer by a bite at the nape of the neck which severs the spinal cord. A young Wild Pig would have been easy prey, especially as the Lynx hunts and the Pig forages for food at night.

The Grey or Timber Wolf in its heyday probably enjoyed a greater range of territory than any other land mammal for it was common across Eurasia, North America and Western Greenland. Living in both open country and woodland it would have been equally at home in the Upper Dale throughout the Great Spring and Summer. It is a ferocious carnivore and may eat as much as one fifth of its own body weight at a sitting, after which it can go without food for a considerable length of time. A lone hunting Wolf is quite capable of bringing down a full grown Deer and when hunting as a pack of up to two dozen animals there is nothing too large for them to tackle. The method of hunting is not to outrun their prey, for they are slow loping movers reaching a maximum of some 45 kilometres per hour only for short bursts. A better term would perhaps be *run-out*, for these large dogs can keep up the chase for hour after hour until eventually their prey tires and they then move in for the kill. Apart from Wild Pig, Deer, Reindeer and Elk their prey may well have included two large members of the Cow family, the Wisent or European Bison and the Aurochs, the ancestors of our domesticated cattle.

The Wisent is very closely related to the Wood Bison of North America and it may well be that it originated in North America, migrating across the then dry Bering Straits to colonise Eurasia. Their main food is bark of which they eat a great

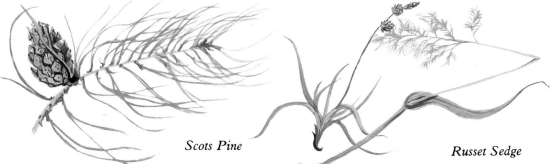

Scots Pine

Russet Sedge

European Bison or Wisent with Honey Fungus on Scots Pine

PLATE 38

variety although specialising in that of Willow, Poplar and Aspen. They will also eat the young shoots of evergreens, Hazel and Oak, rounding off their autumn diet on a surfeit of acorns. If you want to see the massive, stately Wisent in its natural surroundings you must make the pilgrimage to the Bialowieza Forest in the east of Poland where a herd still lives. This beast was miraculously saved from extinction by rebreeding and re-introduction from some thirty individuals that remained in Zoos after it had been wiped out in the wild by the aftermaths of the First World War. There you may see, if you are very lucky, Wild Pig, Wisent and Wolf living in some of the best natural forest left in Europe.

The Auroch was not so lucky, for it passed into extinction in a park in Poland in 1627. The bull stood six and a half feet at the shoulder and had long upcurving horns, a black coat with a white stripe down the centre, white curly hair between the horns and a grey-white muzzle. Our knowledge of its habits stems from modern animals which have been recreated by back breeding from several lines of modern cattle. The successfully reconstituted Auroch not only looks like its forebear but is extremely agile and very aggressive, traits which must have been selected and bred out as Man took their ancestors into captivity and tamed them for life on the farm.

All these and many more lived on the bounty and survived within the protection of the summer forests, each one, whether green producer, herbivore, carnivore, omnivore or decomposer, playing its own part in the cycling of minerals.

Remember, all living things, and that goes for you and me, are in essence no more than that, minerals borrowed temporarily from the environment to which they will return as death and decay takes its toll. Our allotted span is but three score years and ten; for those 'lucky' few perhaps a little more. Like all living things we sense eternity by playing a part in the greater whole, the living landscapes on which we so depend.

Walk through a forest at any season and sense that living continuity. The soft fall of autumn leaves that are absorbed into the goodness of the soil throughout the sleep of winter, releasing minerals to feed the new growth of spring, while in the fullness of summer, water, itself a raw material of the process of production, pours up the trunks, carrying vital nutrients and by evaporating from their surface cools the working leaves. Sense the phytomass, those incredible hulks, trees, bursting their bark and holding aloft green canopies which challenge the power of the sun itself. Their presence is so overwhelming that on entering a forest you blink almost blind until your human eyes accommodate to their gentle giant-like presence. Even on the most blustery of days, the forest floor remains unruffled, calm; and when it rains the canopy of leaves protects the living soil from direct conflict with the erosive power of rain drops. The mass of branches funnels the life giving water down to the trunk and on, to flow and disappear amongst the roots anchored in the well structured forest soil. Watch as a shower of rain revitalises the denizens of the forest floor, Herb Paris, Yellow Star of Bethlehem, Wild Garlic, Dog's Mercury and so many more, and as you do, remember that just as there is structure above your head, so too the structured life beneath your feet reaches down, following the water on its way to hold, bind and recycle those minerals which otherwise could be lost to the detriment of the whole living system.

Oh, to have walked the forests of the Upper Dale back in those days of high summer when the average annual temperature of that place and the rest of Britain was some 2°C higher than it is today! In stark statistics such a difference seems of little significance but that is all it took to clothe the Upper Dale in sylvan majesty.

A firm date of 4,000 ± 60 years bc shows Man the hunter in the Upper Dale. His flint tools have been found in association with the bones of Aurochs. Only a hundred years later a pollen grain of a cereal found its way into one of the living history books of the Upper Dale hinting at the shape of things to come. At that time, around the 550 metre contour, Hazel, Elm, Oak, Alder and Birch in that order held sway upon the fell. The scene of the chase and of the kill can thus be readily envisaged, and note the presence of Alder a newcomer on the scene.

Alnus glutinosa is, when mature, a rather untidy tree with a mass of branches which are at first upswept and then level out to produce a broad

EUROPEAN WOLF

PLATE 39

pyramidal crown casting dense shade. It bears its separate male and female catkins before the leaves unfurl, and very handsome they are, the former long and drooping, the latter oval and erect, and both start off a lurid pink/purple in colour. The male catkins open to release a shower of yellow pollen while the female ripen to form a green egg shaped cone which becomes woody with age and remains on the trees.

It is a plant of damp soils and if the conditions are right it is a very aggressive and successful invader, in part because its roots harbour very special bacteria. These act not unlike those which produce the Witches' Broom on Birch, in this case causing the roots of the host plant to branch profusely, the short stubby branches swelling to accommodate the bacterial populations within and the whole mass ends up looking like one of those pomanders studded with cloves. This is a case of symbiosis, mutual help between two organisms: the Alder provides the bacteria with board and lodging, the bacteria in turn provide the tree with a ready source of nitrate fertiliser. The bacteria in question have the ability, along with those blue-green algae which were amongst the first denizens of the Upper Dale, to fix nitrogen from the atmosphere.

It is well worth looking around the base of an Alder to see these singular masses or root nodules part submerged, peeping from the surface of the wet soil. Better still, where Alder grows alongside a stream it is possible to find strings of the nodules each washed clean and trailing out in the current.

With such built in advantages Alder will oust all competition and it is interesting to see that as it came to the Upper Dale in force, so Willow almost disappeared from certain sections of the landscape. As Alder asserted its self fertilising presence, ousting Willow from the wettest streamside locations, Pine also began to show a marked decline. This change-over in the character of the forest is so much in step that it seems safe to infer that it must, at least in part, have been due to a direct replacement, Alder beginning to grow on land originally occupied by Pine.

The Coming of Man

It was into this warm, well watered, wooded landscape that Man made his first significant excursions into the Upper Dale to hunt and gather in the richness of the summer scene. There is no reason to doubt that early cultures made short visits into the area from their strongholds further south and from the coastal landfalls which were established in mesolithic times. The coastal sites show an evolution through time, the early phases being marked by small and relatively simple flint tools; for example the 'pygmy' points or triangles often not more than half an inch in length but beautifully and delicately shaped by secondary chipping along two or three edges. These 'pygmy' flints were fashioned from an oval pebble and they were used as harpoon barbs and small arrow heads. Similar flint cultures are known from Belgium and Holland and it must be remembered that at the time much of the southern North Sea was dry, itself covered by lush forests, remains of which are known from below present low water mark at several points along the coast. It would seem that these early visitors hunted only small animals supplementing their diet with fish and shell fish; one reason why they kept close along the coast and probably even there were only summer visitors.

In time the early mesolithic tools were replaced by much larger and more advanced flint implements the workmanship of which was rougher. These include larger blades which are less frequently sharpened with secondary chipping, large scrapers and definite arrow heads, the latter shaped from a thick piece of flint that has been chipped down from both sides to form a bi-convex section and then worked into a distinctive leaf shaped form.

It appears that some of this new technology in flint was introduced by other cultures and though there is no indication as yet from whence they came, the evidence shows that they not only settled on the coast but also moved inland along the rivers and into the Upper Dale, the culture evolving as they went to produce arrow heads with flat and even concave bases. As these hunters moved inland they must have come more and more to depend on larger prey and less and less on the easy pickings of the sea, estuary and lowland rivers and so the flint tools they left

MESOLITHIC MAN

PLATE 40

behind changed in character. There is, however, ample proof that even within the Dale inland colonists led a dual existence, their lives divided between the richer lowlands of the main river valley and the fells, including those of the Upper Dale. In this way they made full use of the potential of a whole range of habitats moulding their nomadic lives to the pulse of the seasons.

Again I am jumping ahead, for the period which spans these cultures also spans another major change which was taking place not only in the Upper Dale but also across the length and breadth of Britain.

To account for the dramatic changes in vegetation which were about to take place an understanding of the nature of the soils of the Upper Dale is of great importance. Although the parent materials of all the soils of the area are the sandstones, shales and limestones of the carboniferous age, together with the intrusive Whin Sill, it must be remembered that the whole area had been worked over by the ice of the last Great Winter. Thus apart from the sharpest outcrops and steepest slopes, much of the area was covered with glacial drift and especially by boulder clay which included material from the whole range of different rocks. This is certainly true up to the 600 metre contour above which there is little or no boulder clay of any sort, it presumably never having been very thick and what there was had been removed by erosion during the early part of the Great Spring before vegetation had added its stablising influence.

Let us suppose that as the climate improved during the early part of the Great Summer the surface layers of the soils developing over boulder clay dried out sufficiently to allow their colonisation by shallow rooting trees. The best candidates for this role would have been Pine and Elm, for both are able to thrive on a mass of superficial roots which can tolerate a certain amount of flooding. However as the climate got wetter the upper layers of these clay based forest soils would tend to remain waterlogged for longer and longer periods, providing conditions ideal for the colonisation of Alder which was moving up the valley bottoms.

It is of interest that many of the living history books also show an increase in the abundance of grass, Meadowsweet, and members of the Buttercup, Daisy, and Rose families from this time on, all of which point to an opening up of the forest cover. Another significant feature is an increase in the presence of members of the Heather family including *Calluna vulgaris* itself from this point up. As none of the Heathers are likely to thrive either as members of the ground flora of broad leaved forest or on lime rich clay soils their presence is best explained by the development of acid peatlands in the area. Depressions away from main drainage lines where the ground water came to the surface and remained more or less stagnant could have become nuclei for the formation of acid peat which would eventually spread to cover much of the Upper Dale.

The time of climatic optimum and the climax of the summer forests was past; the Upper Dale was as productive as it would ever be. Far away to the south the English Channel was rapidly filling with water, closing the way for the natural colonisation of more land living organisms, unless they had the power of flight or skill of building boats. Britain stood apart, an island swathed in a changing pattern of forests mature and past their prime. The change-over from the warm dry Boreal to the warm wetter Atlantic period lasted some thousand years and as it did Alder continued to capitalise on its soggy bridgehead in the Dale, Hazel, at least in places, making a distinct comeback. Then around 2,900 years bc the next major change took place: Elm showed a marked and rapid decline, almost disappearing from the Upper Dale.

Close along the main river the combined effects of the increased rainfall draining through the Upper Dale acted as an agent of erosion, slowing the rate of peat formation and in places sweeping some of the record clear away. Perhaps if these valley bottom history books had been left intact we would have more data concerning the decline of *Ulmus glabra*, a decline which was going on apace across the face of Europe, providing the annals of palynology with one of its most vexing problems, not to date, but to explain.

Certainly the climate continued in its wetter mood and Alder continued its expansion within the Dale. The shallow rooted Elm could thus have followed the same fate as Pine, 'flooded' out from the clay based soils. However, if that were the

European Brown Bear Cub

Trout

Purple Saxifrage

Sandwort

Marsh Horsetail

Water Forget-me-not

PLATE 41

case it might be expected that Elm would first disappear from the higher wetter altitudes leaving Oak and Hazel to dominate the upper catchments. The reverse was in fact the case, for Elm pollen remains firm in the living history books of the Upper Dale for almost 500 years after it showed its main decline in the lower landscapes.

Dutch Elm disease, that scourge which is again leaving its distinctive trail of destruction across the face of Europe, could well have accounted for the Elm decline: an attack of plague proportions slowly moving up the Dale, its progress perhaps held back by the cooler upland climate. Likewise the gradual spread of Man slowly moving through the Dale would certainly tie up with the chronology of Elm decline within the area. The main problem with the latter explanation is whether there would have been sufficient men about at that time to have such an effect on the living landscape and why was it only Elm that suffered?

The first men in the Upper Dale were without doubt migrant visitors gathering and hunting where and what they could. They were thus part and parcel of the living landscape playing a role within the climax forest much like any other large omnivore. Another hunter gatherer was also present in the Dale at that time and remained until at least the tenth century AD; his name, *Ursus ursus*, the Brown Bear. When it comes to food Bears, like people, are funny things, for though they are classified among the carnivora (meat eaters) they display a cosmopolitan palate, enjoying fruits including those of *Arctostaphylos uvae ursi* (Bear Berry) which still grows abundantly in certain parts of the Dale, honey, grubs of bees, many other insects, small animals in great variety, fish and even young Deer. They, like Man, are able both to move about and to change diet, taking advantage of whatever the landscapes have on offer.

Meadows in the Sky

The second wave of human influence within the Upper Dale brought with it new technologies which were going to have drastic effects on those living landscapes.

The culture was Neolithic and it was spreading with great speed northwards across the continent. The advanced technology included axes of polished stone and a knowledge of animal husbandry. With the former men could cut down trees and with the help of fire had it in their power to clear considerable areas of land. With the latter they could ensure a continuous supply of meat to supplement their hunting. The overall result was that they began to manipulate the living landscapes to their own end, changing the natural patterns of maturity which had evolved over thousands of years. They were still dependent on the fertility of the soils which had developed through the same span of time but they now began to divorce themselves from the natural patterns of production, imposing their will on the evolved maturity of the summer landscapes.

Whether these Neolithic people brought domesticated animals with them or tamed the local Aurochs is impossible to say, although I feel that the latter is more likely as movement through the forested landscapes of those times would have been made even more difficult by the encumbrance of herds of animals. The route into the bounty of the Dale was probably from the Pennine Ridge, for at least some of their polished stone artefacts were from the Langdale factories situated far to the west. Thus the first 'trade' routes were opened up and Man became a geomorphological force scarring the face of Mother Earth herself and carrying the products to be littered in another place.

Whatever the real answer concerning the origin of their stock, the main problem which faced the Neolithic herdsmen was how to feed the animals in a wooded landscape. In the absence of open grasslands the foliage of the trees was one obvious source of fodder and with Langdale axes to hand they could have quickly polished off the job, bringing the aerial feast down to ground level. Likewise leafy twigs could be safely gathered in and stored for winter use.

It is well documented that such practices of forest husbandry are carried on to this day in some of the remoter parts of the world. I was, however, somewhat surprised to find, not far off the beaten tourist track in northern Spain, that the same practices continue to this day. I was more than fortunate on two occasions to

Lady's Slipper

Goshawk with Aurochs

PLATE 42

live in and around this community of gentle upland farmers and learn their ways. Despite the encroachment of the pressures of the twentieth century it was an idyllic existence, at least in the warmth of summer. My favourite pastime was sitting in the late afternoon stripping leaves to store for fodder and twigs to be used as kindling wood throughout the winter. The cattle stood all around waiting patiently for any stray leaves which came their way: the conversation always concerned the weather and the prospects of harsh winters to come.

The leaves in question were those of Elm and Aspen, not because they were the only trees in the vicinity but because they were the ones preferred by the animals. If you want to put leaf preference to the test yourself then go collect a handful of leaves from a whole range of different trees; mix them well and place the whole mass into a hairnet. Bury the hairnet just below the surface of the soil and mark it well so that you can inspect its contents at regular intervals. When you do you will find which leaves are most palatable to the denizens of the soil, for they will disappear first.

Among the first to go will be the leaves of Elm for they are rich in both protein and carbohydrate. Furthermore, of all the trees of the summer forests, Elm provides the densest canopy, which means most leaves per unit area of the ground, one mature tree being equivalent to two acres of ungrazed meadowland. They were thus the richest of the meadows in the sky.

One well grown Elm felled in its prime represents in crudest terms 20,000 million joules of energy. If the farmer was lucky 100 million joules, the energy of the leaves and buds, would pass to his cattle, the rest to cook his food, warm his home or to the decomposers of the now open soil. The total mineral content which had been locked up in the tree, in kilograms, is 7 of calcium, 3 of nitrogen, 2 of potassium and 0.3 of phosphorous. This would be flooded onto the wet soils and in the absence of the living structure of the forest much would be lost to the erosive power of the rain.

Translate this picture of destruction from one individual Elm to the whole forest economics of the Upper Dale. Four million tonnes of standing crop with an

annual production of some 140,000 tonnes which included 60,000 tonnes of
assorted leaves. The mineral budget in round terms of metric tonnes (each one a
thousand kilogrammes) is calcium 35,000, nitrogen 10,900, potassium 6,150 and
phosphorous 955. Add to this the stores present in the rooting horizons of the soil,
again in metric tonnes; humus 18,000, potassium 1,700,000, calcium 749,000,
nitrogen 282,000, phosphorous 63,400.

All the figures ar but 'guesstimates' based on values for contemporary forests
growing under climatic regimes which are similar to those of the Upper Dale
during the Great Summer. However crude they may be they give some indication
of the potential created over some six thousand years of soil formation and more
than four thousand of forest evolution, enormous potential, much of which was
about to be lost. The evolved economy based on the retention and recycling of
nutrients was smashed. The nutrient stores which had been held in available form
began to bleed away. The soils washed clean of much of their potential lost heart;
the end of the summer forests was in sight.

It was not just the biomass of the forest and the stores of minerals that were
lost but also the rich diversity of forest life, countless populations of animals and
plants which found safe sites only within the continuity of the forest cycles. It did
not all happen at once for it took Man and the changing climate, working in unholy
alliance, almost four thousand years to complete the destruction.

But all was not lost for the rule of nature is unbending: wherever there is
potential it will be used and so the life in the river must have benefitted from this
long term influx of extra nutrients. The algae, mosses and other plants of the river
and its banks must have grown more rapidly and the Bullheads, Brown Trout and
Sticklebacks followed suit. Lower in the river the great Salmon and Sea Trout
must have swum through more productive waters as they came each year to spawn
below the great waterfall that barred their way to the upper reaches.

As the forest soils became leached of their minerals they changed in character, the upper horizons becoming more acid while lower down redeposition of iron
and other minerals formed what is known as a hard pan. These layers of mineral

*Wild Cat with Wild Cherry, Star of Bethlehem, Adder's Tongue
and Speckled Wood Butterfly*

PLATE 43

material, cemented together by humic products washed down from above, further impeded drainage so that the soil became wetter. Over large stretches of the Upper Dale this further antagonised the new climatic regime in which precipitation always outstripped evaporation over much of the year and the condition for the growth of acid peat became widespread, centred at first on those depressions in which it had already begun to form. Over much of the area the new peat forming plant communities were dominated by the Common Reed and by numerous Sedges, a last gasp of productivity before the blanket of acid peat itself came to dominate the landscape.

If you know your topography and know the signs it is possible to pick out such places and, plunging your arm down through the soft peat, you may be rewarded with the shining silver-white sheaths of Reed rhizome, flattened but still recognisable. These, together with the fruits of Bottle Sedge and mats of *Drepanocladus fluitans* (a moss) and seeds of Bog Bean mark the final passing of a more productive age.

To be used in Evidence

The living books of history were moving into their most expansive phase just in time to take down the evidence of massive and important changes in the Upper Dale and the chief chroniclers of this phase of change were the Bog Mosses or *Sphagna*.

Like all mosses the leaves of the *Sphagna* are only one cell thick and yet they are constructed in such a way that they have an almost spongy texture. Each one is made up of small elongate cells which are alive and bursting with chlorophyll. These are sandwiched between much larger cells which are dead, empty, and what is most important, are furnished with large pores which let the water in, and bands of strengthening which hold the cells open ready to receive it. The result is that the leaves act just like a sponge and each one of these tiny plants can hold many times its own weight of water. Thus as the plant grows upwards it carries with it its own supply of water and as these tiny plants grow, appressed to each other in dense mats, the whole acts like a sponge carpet carrying the surface water table up: lovely stagnant water ideal for the formation of peat and that is exactly what happened over much of the Upper Fells. Conditions became right for the growth of the *Sphagna* and as they grew their remains did not rot but became sub-fossilised in the form of peat. So excellent was their state of preservation that it is easy to identify which of the various species of bog moss were most important in changing the face of the fells. The bulk of the lower peats are made of *Sphagnum fuscum and S. imbricatum* and from this time hence the record in the peat is both detailed and expansive.

Excellent as a layer of peat may be at recording the changes in a landscape it is itself a major perpetrator of change, for peat slowly but surely shuts off the plants growing on its surface from the mineral supplies in the soil below. This is of no consequence to the majority of the bog mosses themselves for their evolution has made them both frugal in their mineral requirements and masters of mineral retention and recycling. Thus after the first flush of reed and sedge dominated peats the pattern settled down to one of acid frugality, a blanket of bog mosses on which few higher plants could thrive; Heather, Cross Leaved Heath, Single and Many Flowered Bog Cotton, Deer Sedge, White Beaked Sedge and Mud Sedge to name the most important among the few. The pink Cyclamen-like flowers of the Cranberry gave colour to the hummocks of Bog Moss in early summer, while the handsome spikes of Bog Ashphodel painted the blanket peats first with the special yellow of their flowers and then with the lurid orange-red of their fruits. This latter plant goes by the Latin name of *Narthecium ossifragum* which means *fragile bones*, hinting at the paucity of calcium in the diet of any animals which attempted to graze an existence from these acid peatlands. Grazing both of wild and domesti-cated animals would thus have concentrated more and more away from these areas putting excessive pressure on the resources of the forests that remained.

Cultivation of crops was by now a fact of life within the Upper Dale for the pollen of cereal, Sorrel and Plantains are a constant feature of many leaves of the living history books. The plots were in all probability quite small and as their

Harvest Mouse on Jacob's Ladder

PLATE 44

fertility decreased new ground would be cleared by axe and fire, and new crops planted. Hunter, gatherer, herdsman and now agriculturalist, the new Dalesmen were evolving, fitted with new skills, and they needed to be, for as their numbers increased it must have been increasingly difficult to wrest a living from the changing landscapes. They were thus forced by circumstances to turn their minds and dextrous fingers to more difficult types of forest and terrain.

Oak, with its legendary hard wood, was more difficult to cut down and Alder grew only on the wettest mineral soils which would be difficult of access by all animals as would the growing blanket of peat. However, around the year 1,500 BC Oak began to go the way of Elm, falling more readily to the ring of Bronze Age axes which were by then in use within the Dale, and as the Oak disappeared grasses came in to take its place. The landscapes of the Upper Dale began to open up and many of the plants of the last Great Spring came out of hiding to beautify the Dale once more.

Miraculously, we know that some of them were there throughout the Great Summer for we have records of their pollen, scattered across the peaty record. Dwarf Birch, Hoary Rockrose, Jacob's Ladder, Mountain Avens, Purple Saxifrage, Sea Plantain, Spring Gentian, Starry Saxifrage, Thrift and Yellow Mountain Saxifrage were all there throughout the period when the fells were first covered with forest and then with peat. Their refugia were in all probability on the sugar limestone, for the records which we have come from a living history book which developed quite near an extensive outcrop of this unique rock type. It is easy to understand why large trees could never find a firm foothold on such a friable deposit. Likewise it seems unlikely that such a freely draining substratum could have ever succumbed to the suffocating blanket of wet peat. Such areas had in all probability remained as more open vegetation throughout the 9,000 years which saw so many changes on the fells and the delicate but hardy plants of spring continued to grow there safe from competition. Likewise the steep cliffs of Whin Sill and other rocks must have formed safe sites for other plants throughout the time and certainly *Woodsia ilvensis*, Holly Fern, Bear Berry, Fir Club Moss, Mealy

Primrose, and Mountain Everlasting may have hung on in those spots. Where the river eroded back through eskers and drumlins of glacial drift neither forest nor bog could get a firm foothold and lime rich ground would have been bared anew each spring as meltwaters swelled the erosive power of the river.

Such places thus provided refugia, islands of open vegetation set within a sea of vegetational change, 'nunataks' on which some of the plants of the Great Spring survived in splendid isolation, waiting their chance.

The time had come: the potential of new grasslands, wet, dry and all shades in between, were there ready to be exploited. Thus it was that some of the beauty of the Great Spring came out of retreat to colour the more sombre landscapes of the Great Autumn. The scene was reset and the living landscapes of the Upper Dale were much the same as those we see today, the upper fells clothed with a mixture of blanket peat, part dominated by Sphagnum and Heather, interspersed, especially on the limestone outcrops, with grasslands rich in species.

IRON AGE WOMAN

PLATE 45

The Great Autumn

Empires Evanescent

F LINT, polished stone, bronze and iron, the products of the ephermeral cultures of Man, had done their worst and by the Late Iron Age the glory of the summer forests was but a memory of another culture's past engrained in the pollen record.

Some 4,000 kilometres to the south east an event took place which, at least in the minds of certain men, gave Time's arrow a fixed point of reference on its unbending path from the past to the present and on into the future. That singular event, the birth of Christ, passed unnoticed in the Upper Dale although, as we shall see, in years to come it was to be both cause for comfort and for conflict within those then pagan landscapes.

Straight roads that led from Rome were at that time pushing both north and west across the broad face of Europe, breaking barriers, linking cultures, creating new history, the sort that is recorded by the hand of Man. It was a new wave of constructive destruction following late in the wake of the melting ice. The Romans brought with them the trappings of an Empire, new strategies of hunting and of war, techniques of farming, methods of building, mechanisms of transport, processes of legislative law and order, systems of export and import and, what is perhaps the most significant of all, the written word.

So the might of Rome, spelled out in living Latin, came to Northumbria but the Upper Dale was little affected. The legionaries marching up the Great North Road passed it by and it became somewhat of a backwater. With improved implements made of iron both the need for wood and the efficiency with which it could be obtained had increased. The growing population of Man had turned its destructive attention to the promise of the dense forests of the lowlands which

97

once tamed could produce much better crops than could be won in the hills, so much so that even before the Romans arrived much of the area had been deforested and turned over to farming by the Britons.

But the Upper Dale was still populated probably by hardy individualists who continued the schizophrenic existence of their ancestors which allowed them to make use of everything the living landscapes still had on offer. Throughout the spring and summer they hunted the valley and the fells, grazing their animals on the grasses and sedges of the higher pastures. The favourite haunts of those animals and of the wild ones which shared the pasturage were, without doubt, along the outcrops of limestone where their diet was varied, if not enhanced, by all the arctic alpine plants. This increased presence of animals must have been of key importance in keeping the process of succession at bay, which could have so soon covered them again at least with Hazel scrub. There also seems but little doubt that the farmers raised crops of cereal on the best soils lower in the valley and there they built their homesteads. The remains of circular huts set in a small stockyard, itself surrounded by a protective embankment, have been found at a number of choice localities. Within these enclosures leaves were stored for winter fodder and the cattle could be sheltered safe from wind, weather, and possible attack by Wolf and the rustling presence of other farmer gatherers. There too some people must have practised their crafts – burning charcoal, forging iron and firing clay.

It is of interest to speculate how many prehistoric farmers may have lived off these autumnal landscapes, and please remember it is but speculation. Comparison with estimates for similar cultures in similar situations suggests a figure of two hundred *Homo sapiens* for the whole of the Upper Dale as not unreasonable.

In comparison with our own lifestyle it must have been a hard existence and study of primitive societies reveals a pattern of part work, part leisure, based around family groups which worked, lived, and learned together.

These hunter farmers may well have been hardy individualists but their iron implements and embanked farmsteads afforded little protection from Roman soldiers. The Romans came, they conquered, but evidently they didn't like what

Dales Pony with Meadow Pipits, Bird's Foot Trefoil and Harebell.

PLATE 46

they saw in the Upper Dale. Perhaps the locals were too individualistic in their approach, perhaps the wet upland environment offered too little to make permanent occupation worthwhile. Who knows? But it is a fact that the Romans never colonised the Upper Dale. They used it as a place to hunt, perhaps to train their men, and as a short cut in the summer north towards Corbridge and the Wall. There is little doubt that these uplands needed constant patrolling, even if it was only to remind the locals of the Roman presence. This was carried out from the nearby forts of Braniacum, Lavatrae, Maglona, Verterea, and Vinovia, where both infantry and cavalry were based. Likewise along Hadrian's Wall with garrisons numbering in the thousands, and less than 100 kilometres to the north the Upper Dale may have provided labourers and sturdy black Dale ponies; food, both meat and corn; skins, hides, wool, tallow; in fact all the products of the living landscape.

Solid evidence of the Roman presence in and about the Upper Dale comes in the form of a host of small artefacts and four altars carved in honour of the god Silvanus, a god of the woods, a benefactor of all hunters. The most famous of these comes from the adjacent dale where it may still be seen, a pagan artefact now held in trust in a Christian church. The clear inscription which had been recut speaks of both pomp and circumstance.

'Sacred to the unconquered god Silvanus, C. Tetius Veturius Micianus, prefect of the *Ala Sebosiana*, freely sets up this in fulfilment of a vow because of the capture of a boar of outstanding size which many of his predecessors could not bag.'

The killing of that Wild Pig did more than boost the ego of the hunter; its carcass fed both his compatriots and his dogs; the recarving of the altar gave employment to a craftsman and the new cut letters a fresh habitat for the growth of microlichens. Wherever there is potential it will be used; this, as we have seen, is the unbending law of nature whatever Empire holds sway at a particular time.

With the Romans came the trade routes spanning the then known world, goods and chattels began to flow in all directions and hunting dogs bred for export may well have cut their teeth on the bones of that Wild Boar.

Roman rule lasted for no more than four centuries but in that time the boundaries of the British scene were linked by straight roads to London, Dover and thence to Rome, and law and order was maintained for much of the period. The living history books tell the story in strict detail although with such a short span of years it is often difficult to trace out actual sequences in time. Clearance continued and as markets expanded farming became the order of the Dale. The cultural pollen grains which are associated with this phase of open-faced stability are those of Wheat and Barley. The horizons are also marked out by both the constant presence and increasing abundance of spores of Bracken (a fern which soon invades abandoned land) and the pollen of Dock and Plantain which are both weeds of fallow land and members of the flora of more permanent pasture. Likewise the pollen grains of the Ribwort Plantain or Waybread, a plant which is both tolerant of grazing and, as its latter name suggests, of trampling, increase in abundance. Thus the new trade routes were in all probability edged with green and the marching feet would have sent clouds of that particular pollen skywards as they went. The way was open and new methods of earning daily bread were just around the corner. Ever since the first mesolithic hunters arrived on the scene bringing with them flints collected from Flamborough Head where it outcrops with chalk as cliffs beside the sea, the Upper Dale was open to Man made imports and exports. Each new culture increased the range of imports and the volume of eroded material exported down the river back to the sea. Throughout the pre-historic period the flow of artefacts and the art of barter must have increased as the Hill People exchanged the excess gleaned from the hills for the goods they needed. With the coming of the Romans a new word came to be used in the exchange and mart of the Dale, and coins bearing Ceasar's image became the currency of trade. So money thus raised its imperial head, and the words richer and poorer took on new meaning as Man made one more decisive step away from the discipline of the seasons. The dual hunter/farmer economy became salted with commerce. Coinage could be carried more easily to and from distant markets and could be stored

SILVANO INVICTO SAC
C TETIVS VETVRIVS MICIA
NVS PRAEF ALAE SEBOSIA
NAE OB APRVM EXIMIAE
FORMAE CAPTVM QVE
MVLTI ANTECES
RES TVS ERA OARE
NON OTVERNT V SLP

Rabbits with Roman Altar

Roman Snail

Herb Robert

Great Plantain

Pedunculate Oak

Lichen

PLATE 47

safe from damage by the vagaries of the natural seasons. Hard cash would in the fullness of time become the common denominator of all transactions.

Another much softer import was soon to make its presence felt within the confines of the Dale, both in a visual and commercial sense. The Romans brought with them a white woolled breed of Sheep to mix and mingle with the local stock of brown woolled Soay Sheep. The latter get their name from the island off the west coast of Scotland where in genetic isolation they still breed true to this day. The resultant mixture of these breeds laid firm foundations for the wool industry of future cultures and other Empires.

The presence of Sheep of any sort is very difficult to ascertain because their bones greatly resemble those of the Goat. In fact to tell the difference even the expert must carry out a great deal of measurement and statistical evaluation. This may seem surprising to all who have only seen the shapely modern breeds for they, even from outward appearance, are as different as chalk and cheese. However, those of you who have been to warmer climes and witnessed less developed forms of agriculture will know how difficult it is to tell the Sheep from the Goats and perhaps will understand the dilemma of the archaeologist.

Goats provide meat, milk and hide: Sheep added wool to the list of commodities, spinning and weaving to the list of crafts and warm clothing to the wardrobes of both the Romans and the Celtic people who learned not only to bear but to benefit from their yoke.

Not only did the 'blood' of the breeds of *Ovis aries* mix and mingle to produce new stock but so too the 'blood' of the conquered and conqueror mixed to produce new families which had hybrid allegiance both to Rome and Britain. So it was that when around AD 400 trouble much nearer home called the Roman legions back to defend the heart of the Empire, some Roman blood was left behind and some British stock, perhaps even from the Upper Dale, went to fight in foreign wars.

Roman rule had brought stability to much of Britain and though conquerors they were welcome allies when it came to defending home lands against invasion by Angles, Saxons and Jutes from across the North Sea and Picts and Scots from north of the Wall. With their withdrawal, for retreat would not be the right word, this more stable period of British history came to an end as new invaders took over the potential of a fallen Empire.

Over much of Britain the living history books continued underwriting the less permanent records of the scholar's stylus and went on to record a resurgence of the forest trees as local systems of stable agriculture broke down under the pressure of new wars. In the Upper Dale and throughout much of North East England it was a full two hundred years before this latter change took place, for apart from local resurgence of Birch the land remained open, much of it under the permanence of blanket peat and heather moor, the rest at the hand of Man and his extensive herds of grazing animals. This survival of the open landscape happens to coincide with the period when Northumbria under its Anglian kings was the 'light of Europe', an oasis of civilisation which bequeathed such masterpieces of art as *The Lindisfarne Gospel* and flourished until its extinction by the Vikings.

Domesday was too Late

In matters which concern the Upper Dale there is a distinct lack of direct documentary evidence until the year 1017, some six hundred years after the stabilising influence of Rome had disappeared from the land. During this time various invasions had come and gone leaving their mark of instability on the land in the form of re-expanding forests. The Upper Dale remained a backwater in these inhuman affairs.

Firm record indicates that the whole area officially came within the dominion of the Bishops of Durham when King Canute himself granted to the Church of St. Cuthbert the lowland township of Standropa together with all its hamlets stretching up valley towards the Upper Dale. This in all probability did no more than ratify or perhaps extend the Church's holdings which probably date back at least to the early ninth century. The name Standropa, which means 'the valley with the stony ground', is of old English derivation and this indicates that at least some of the villages and hamlets as we know them had long been features of the living

Pearl Bordered
Fritillary

Curlew with Chick and Egg

Spring Gentian

Rock
Violet

Bank Hair Moss

Moss and Cross Leaved Heath

PLATE 48

landscapes. Likewise this official link of ownership between the lowlands and the uplands carries on the even older tradition set up by the mesolithic hunters, of mobile *Homo sapiens*, deriving benefit from the full range of environments and products the living landscapes had on offer.

It is of importance to note that the Viking raids which eventually put Canute on the throne of England were, in all probability, at least in part due to overpopulation in Scandinavia. The raiders were in the main young men who left their overcrowded home fields and traditional fishing grounds to seek new opportunities in other lands. Unlike those of the Lemmings, their explosive migrations were not barred by river, lake or sea, and their seeking for new territory took them to Iceland, Greenland and beyond, even to discover a whole new world which they called Vinland. Nearer home a Viking Earl by the name of Rollo founded the Dukedom of Normandy in 1011 and it was his descendant who conquered England in 1066 and set it firm on the road to feudalism.

The feudal system was one of land tenure in return for service and it gradually developed across Europe after the fall of the Roman Empire. Its roots, however, go back much further than that linking men firmly to the area of land from which they gain their livelihood.

We have seen that with the passing of the Romans organised agriculture declined in many parts of England and men were pushed back from dependence on agriculture more towards the life of the hunter gatherer.

Feudalism changed this: indeed it made hunting the privilege of the king and the new nobility, as anyone who remembers the stories of Robin Hood must know. The manorial system, with land held in a hierarchy of fiefdoms and ultimately from the king, with agricultural workers bound to the soil, brought organised and very much improved farming including the rotation of crops. Though one would not wish to push a comparison of this human organisation with a natural order too far, it seems right at this juncture to point to certain similarities between this feudalism and the forest systems which had ruled the Upper Dale for so long. The trees appear to be lords, masters of the society of plants. They clearly

provide protection and a multitude of niches, mini-environments in which many
other plants and animals find safe sites in which to thrive. A hierarchy of strata,
trees, shrubs, herbs, mosses and liverworts, a multidecked solar cell, all together
making best use of the energy of the sun. Yet many of these protectors and
providers, these lordly trees, are dependant upon insects for pollination, animals
for seed transport and the host of decomposers living in the soil for the breakdown
of waste and recycling of raw materials. Together the forest community can
thrive: alone the individuals could not last for long.

The benefits and the constraints of feudalism came late to the Dale. The
intransigent people of Northumbria revolted against the Conqueror and after the
destruction of his garrison at York he harried the countryside from the River Aire
to the River Tees with merciless ferocity, burning buildings and all but exter-
minating the people so that when the clerks of the Domesday inquisition came to
Yorkshire they reported most of it as 'waste' and did not bother with Durham and
Northumberland. The Scots too had a share in retarding development and laying
waste the countryside throughout the border region.

The Domesday survey which records so much about the life and economy of
other parts of England comes too late to tell us about conditions in the Upper Dale
in the eleventh century.

We can only guess how many men made use of the living landscapes of the
Upper Dale. The forests once more became important features, protected territories in which the nobility could hunt, be it for pleasure or necessity. The Upper
Dale came under forest law which included the creation of game sanctuaries still
remembered in the local name Frith, which means a place where the Deer may
bear their young in peace. This in itself betokens a change of attitude in the
relationship between Man and the living landscape, the first real hint of the
problems of relating resources and resourcefulness. Though the taking of game by
the common people was curtailed, they carried on their husbandry within the
reserves, as they still do. There is thus little doubt that the Red Deer shared their
grazings with cattle, Sheep and Goats and with less domesticated stock such as the

Red Deer with Juniper

PLATE 49

Rabbits which had been introduced, in all probability, by the Saxons.

The local lords collected their aids and reliefs, hunted the Deer and perhaps the local races of Brown Bear to extinction. Wild Aurochs had been wiped off the face of the Upper Dale by the early Iron Age, and the Romans had seen the last of the Elk. No wonder then that the stocks of Red Deer with their enormous spread of antlers were conserved by the new forest laws.

The first documentary evidence of a building within the Upper Dale comes with a grant allowing the construction of a house 'at the head of Kavaset next to Ethergilebec'. The instructions were both detailed and clear as to the size of the building, 20 × 2 perches with five acres of land enclosed by hedges and ditches, and to their use, the winter pasturage of mares and their colts. The grant was to the Abbey of Rievaulx and it supplemented rights granted at the same time to graze sixty mares and their offspring throughout the forests of the Upper Dale.

There seems little doubt that this substantial building simply added to a scatter of farms which already existed in clearings throughout the Upper Dale, a pattern which had probably remained almost unchanged for more than a thousand years. To prove, or at least drive home the point, it is as well to note that this building was within two kilometres of those desirable residences overlooking the great waterfall which were probably built in Iron Age times, and were excavated by Dennis Coggins who lives today in a house which stands midway between the two.

From this time on the written records are more numerous and more exact, telling of changes in overlordship, of relations with Church and state and, in 1218, the first mention of the excessive demands of the lead and silver mines for fuel from the forests. Conservation began in earnest with the closing of the forests of the Upper Dale. Thus, 750 years ago the conflict between industry and nature conservation reared a stubborn head which would be raised in far more serious conflict in the twentieth century.

The Crusades, Wars of the Roses, Reformation, and many other happenings came, while the living landscapes of the Upper Dale changed but little; so too the way of life of the common people who made their living from those landscapes.

Unfortunately little is known of these 'lesser' classes and one can only guess that their fortunes must have fluctuated along with those of their lords, buffered, at least in part, by the productivity of their holdings within and beyond the confines of the forest. In good years there would have been more than sufficient corn, hemp, flax, hay, wool, honey and so on, for their own needs and to pay their aids, reliefs, tithes, and taxes. In bad years when late springs, wet summers and harsh winters sapped at the promise of their lands, survival became the only drive of life and new ways were sought to make a living.

The feudal system came, had its day and began to pass away. The reasons for its demise were many, but perhaps most important, and the one which concerns us here, was the increased availability of coinage. Despite the introduction of minted money into the Upper Dale by the Romans (and there have been a number of finds to prove it) the main currency had remained muscle, the power to work for your livelihood, pay your taxes and produce a surplus which could be exchanged for other things.

Both muscle and goods are forms of available energy and as such they can't be stored for long. In contrast, money, whatever effigy it bears, can be saved, stored and hoarded to be set against a rainy summer or a hard winter, or used to pay other men and women to do your bidding or to help you out of trouble.

The vocation of a vassal which had evolved to be a protected meal-ticket now became a price tag for labour with opportunity for middle men to take a share, fair or otherwise. Cash flow, rather than direct exchange, opened up new possibilities within society creating the potential of new livelihoods; places for people who neither owned land and died in its defence, nor worked land and died in the poverty of harsh seasons, but who were paid for other duties and for expertise. Above all, money made way for the growth of specialism, opening up new lines of energy flow and opportunities for different ways of livelihood.

It is not out of place once more to draw attention to a natural analogy. Forest systems store energy in the form of phytomass, mainly wood and humus, which

Bell Flower

Northern Eggar Moth

Valerian

European Brown Bear and Cub with Wild Bees

PLATE 50

provide a buffer against much of the fluctuations of a seasonal climate, a structure within which many specialist organisms, both plant and animal, find a way of life. Likewise, within any natural landscape a mature forest also provides a focal point of living, a reservoir of stored energy which is there to be used whenever it is necessary. Herbivores, omnivores, carnivores alike find shelter and safe sites, for in the forest there are many strategies of protection. From there they range out to feast upon the short term energy flow economies of grass and moorland where the bulk of annual production is put into cycle and little into store. So within any living landscape there is a flow of energy from the less structured, less mature, to the more structured, more mature elements of that landscape.

Energy stored in the form of cash began to play a similar role in the maturing society of Man. The forests began to lose their importance as more and more men forsook their roots in the good English soil and found new ways of industry. If feudalism lingered on, then it did so in backwaters, such as the Upper Dale, and certainly the forests maintained a central, stabilising presence for at least another century.

So it was that in 1670 Roger Baynbrigg argued his claim to the rank of gentleman by citing the fact that his ancestors, who had lived in the same house since at least 1519, had been 'rangers of the forest'. The location of the house is in all probability the same as that at Kavaset, standing within a plot that had been enclosed 500 years before.

The process of enclosure did not happen all at once: it had in fact been going on since Iron Age times, gradually nibbling away at the forest. Indeed it was the natural flow of energy and raw materials towards the forest system which made the act of enclosure, or perhaps the correct term should be exclosure (shutting out), necessary. Likewise it was the importance of the mature forest in the overall feudal economy of the Upper Dale that kept enclosures to an absolute minimum. The lord's right to hunt and his vassal's right to collect wood and gain some benefit from the spoils of the chase were thus maintained.

But times were changing fast and new economies were in the making, economies which would bring about radical changes in the pattern of life not only within the Upper Dale but out across the world of Man. The population was growing, its social economy based, at least in part, on coal mines and manufactories, the latter developing from the workshops of Tudor times and the former a legacy of energy left over from a more affluent past when this section of land basked beneath a tropical sun.

Man was beginning to take his most decisive step away from dependence on the current account of solar energy towards a fossil fuel economy, Coal, oil and natural gas are products of past millenia of photosynthetic affluence, a vast world deposit account of energy and chemicals, there to be used by any organism with the ability to wrest them from the ground.

I now come to a hinge in the story of the Upper Dale. Man, himself a product of evolution, was exploiting every available potential without restraint or responsibility for the future, and with the cash incentive everything seemed possible. The Industrial Revolution was only one discovery away.

Economies which had been based on land holding and working metamorphosed into economics based on the volume of exports and imports; and the flow was as always positively in the direction of the richer system. London held the power and as by the sixteenth century much of its own local forests had been destroyed coal was being imported by ship from North East England to supply its needs. London's need for coal thus became a spur to development in the North East lowlands and the increasing population who worked in the mines and built the new pit villages and towns needed food and clothing; and what is more they had money with which to purchase it. These new, richer communities themselves became focal points in a cash flow economy, new opportunities which could not be missed.

The people of the Upper Dale responded by producing cash crops – cattle, sheep and horses all of which were easy to export on the hoof. The small plots or intakes set on the best ground were only large enough to produce sufficient food for the local families and their creditors and so they had to be enlarged.

COMMON BUZZARD

PLATE 51

I like the term 'intake' for it means just that, an area annexed from the forest and taken into the care of Man to provide him and his family with their needs. It speaks of that special, gentle relationship between Man and the soil, the mediators of which are the members of the plant kingdom. Resource, producer and product, with a hint at the fact that while the land was maintained with care it would remain a productive, viable entity.

Take heed that the end of the Great Winter had seen every square metre of the Dale left with a certain potential for life, a certain amount of useful chemicals held in store within the parent materials that would go to form the forest soils. The Great Summer forests had developed that potential to the full and so had remained dominant features, focal points of the living landscape, until the integrity of their patterns of maturity, based on the recycling of all limiting resources, had been destroyed. The advent of Man and wetter weather had seen much of that potential bleed away downriver as tracts of forest were replaced by wet unproductive moorland which itself stored energy in the form of peat and grasslands which set the pattern of annual production and utilisation. Once again similar massive changes were underway: what would be their outcome?

The small plots which had produced a variety of crops were in the main kept to feed the family, becoming the homefields which may still be seen, green jewels within the more sombre verdure of the 'inbyelands'. Forest was cleared with axe and fire, and new fences and hedgerows came into being, the newly enclosed land being put to grass to feed the animals. The success of these new ventures depended on a good crop of hay which could be stored for winter use. It must have been a precarious existence, for a late winter could spell disaster on a massive scale and yet one presumes that it offered more than the hand-to-mouth existence of subsistence agriculture eked out by hunting in the forest. Money within the grasp of dextrous human hands is perhaps worth more than promises of next year's crops which are in the hands of soil and the weather.

Today, as I write this account there is snow on the ground outside my window. It is March 24th and spring has officially begun. Yesterday I was out on the high fells of the Upper Dale with a farmer who was taking food to his snow bound sheep. All the gas-guzzling paraphernalia of modern farming were of no avail: we had to climb the drifts piled against the enclosure lines and carry the bales by the willing power of human muscle. Each bale smelt sweet, emanating not only the fragrance of last summer but of plants drawn from the whole history of the living Dale which spans no more nor less than the Four Great Seasons.

Amongst the haulms and leaves of Crested Dog's Tail, Red Fescue, Tall Oat Grass and many others were the unmistakable remains of Yellow Rattle, Mountain Pansy, Wood Cransbill, Fragrant Orchid and a whole gaggle of Lady's Mantles and Hawk-Bits, Weeds and Beards. It was fascinating to sit in the shelter of one of these bales of history and dream of the productivity of ten thousand summers. The adjacent bale broke open as the hardy Swaledale Sheep came to feed on the sweet store of energy, revelling in Globe Flower, Bistort, Melancholy Thistle, Red Rattle and Grass of Parnassus. This one had evidently come from a wetter area close along the banks of the river and there to prove it were a scatter of leaves of Tea Leaved Willow and Shrubby Cinquefoil.

I sat entranced as the Sheep pushed to get their share of food and shelter and thought back to 1670 when one John Ray recorded for posterity the fact that Shrubby Cinquefoil grew within the Dale. John Ray was the father of English botany, one of those people whose expertise was able to flower within the new cash flow economies of the seventeenth century which saw the publication of the first three British Floras, *PHYTOLOGIA BRITTANICA* by William How MD, 1650; *PINAX RERUM NATURALUM BRITANNICARUM* by Christopher Merrett, MD, 1667, and Ray's *CATALOGUS PLANTARUM ANGLIAE, 1670*.

Ray's Catalogue gave us the first record of the botanical treasures of the Upper Dale and it was made possible by the new affluence based partly on the use of fossil fuels. Another product of that new affluence was all around me, hay from those meadows which have been maintained ever since by the hard work and loving care of the hardy people of the fells who have kept

David Bellamy with Swaledale Sheep and Shrubby Cinquefoil

PLATE 52

their rights and roots of seasonal livlihood within the living landscapes of the Upper Dale.

I picked at a bunch of hay and found to my delight the remains of a shrubby Cinquefoil flower which took my mind back, beyond Ray, to those first enclosures close beside the great waterfall where Man had first come to live and about which this plant flowers in abundance each year. I thought too of those farming families who have cared for those same lands and thus, inadvertently, for this special plant since baptismal records began in the local parish register.

May 17th (when Potentilla should have been in flower) 1579, William son of Thomas Hutchinson.

November 18th 1584, Roger son of Henry Newby.

April 29th 1638, Rebecca daughter of Thomas Robinson. (The Robinsons lived there for at least 250 years.)

We know that a farmhouse was built upon the spot before 1579 and that several half-crown pieces of Charles I were hidden beneath its foundations, probably in 1648 when Cromwell was advancing up the Dale, for they came to light in the rebuilding in 1838. The old house was oak timbered, thatched, with high gables, low side walls and mullioned windows. Its outer door was of heavy oak studded with iron nails, strongly hinged and secured by a stout oaken beam slotted behind it into grooves in the wall on both sides of the doorway.

One can only wonder whether Rebecca Robinson and her sister Margaret, who was baptised on November 14th, 1641, saw a Mr T. Willisel wandering the banks of the river when he collected the specimen of *Potentilla fruticosa* which he sent to Ray for official record. The name used then was *Pentaphylloides fruticosa*.

Whether or not the Robinsons saw him is irrelevant; what is important is the continued presence of these farming families who looked after the meadows along the margins of which this plant still thrives.

The pattern of management had been set way back in 1131, when grazing of the unenclosed areas (outbyelands) of the forest and beyond had been banned between November 11th and April 1st each year. Between these dates the animals were kept to the inbyeland and fed on hay. Thus it was that the plants of the meadows were allowed to grow ungrazed throughout much of the spring and summer and so they could bear flower, fruit and seed. During this period of growth at least one crop of hay was taken, the sure slow action of the scythe ensuring that the ripe seed was broadcast far and wide. The stooks and bales, once dry, were stacked in the centre of the fields ready for the winter when the cattle and Sheep were brought down to graze, first upon the regrowth and then upon what had been held in store.

This practice of foddering the animals in the meadows in which the plants had grown ensured recycling of those all-important minerals which, in the absence of the forest trees, would soon have been leached away in this wet upland climate, and maintained both the productivity and so the practice. Likewise the use of the scythe and the field stack, the location of which was moved each year, ensured maintenance of the diversity of plants which included among their numbers nitrogen fixers like Zig Zag and Red Clover, Meadow Vetchling and a number of vetches.

A new productive, diverse vegetation of great beauty thus evolved under the caring hand of Man, its plants drawn from the ground flora of the forest it replaced, with others from the natural grasslands of the river edge and continually eroding flood plains, and still others brought down from the grasslands and even from further afield by the seasonal movements of animals.

The only losses from these new stable systems of production were in the meat, hide, wool and bone, which flowed to the richer population of the lowlands. It was a small loss which the natural processes of weathering and deposition could at least in part replenish. The return was cash which could be used to purchase manufactured goods and to start new ways of life.

Lead to Mercury

The Industrial Revolution was a time of massive change arising in part from the discovery by Thomas Newcomen of how to make steam power machines, and steam was produced by fire for which the main resource was coal. Factories and

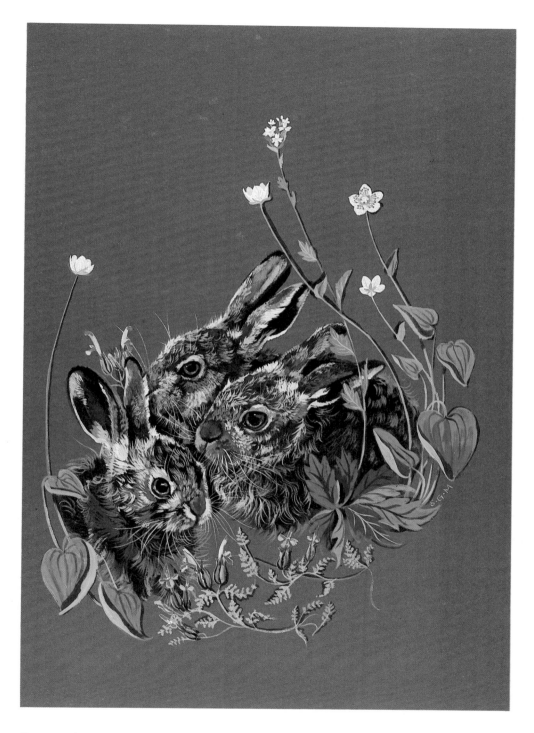

*Leverets (young Hares) with Grass of Parnassus, Lousewort
and Hoary Whitlow Grass.*

PLATE 53

factory towns sprung up around the coalfields and the increasing populations had
to be fed. The flow of energy as meat and agricultural produce from the Upper
Dale down to the lowlands thus increased, but much more important than that
other treasures of the Upper Dale came into great demand.

Deep beneath the parent materials which had formed the productive soils
that had supported Man so well in the past were deposits of minerals which
became of more value as industry diversified. One of these was lead, noted for
both its malleability and ductility, properties which made it ideal for a whole range
of uses in buildings and industry. Many of the rich deposits further south in
Derbyshire had already been worked out and men with mining skills were looking
for new potential in other areas.

To understand the problems and fortunes of mining in the Upper Dale we
must return to the time when the rocky ramparts of the landscape were being
formed. The limestones are the remaining sediments formed in shallow seas which
were subject to constant change, shallowing, uplift, and even drying out. Later
movements of the earth's crust caused the limestone strata to crack and fault; and
it was into these extensive cracks that hot liquids containing an abundance of
minerals found their way. Some of these were volcanic in origin while others were
recirculation fluids derived from deep down within the upper strata. The dead
ends of the cracks and crevices thus became reservoirs filled with rich deposits of
minerals. Galena containing lead, Sphalerite containing zinc, Flourspar (calcium
fluoride) and Barytes (barium sulphate), all there, ready to be mined by anyone
who had the tenacity and the know-how. Against such a background it is easy to
see in the mineral mining business that luck was of almost equal importance to
knowledge, for the richness and extent of each vein was in the main due to the
random processes of flow, directed in part by gravity.

Coal mining is a much less chancy business for a coal seam represents strata
laid down over a long period of time and usually over large areas, so once a seam is
discovered and a shaft is sunk there is much more likelihood of long term
production and employment, though even in the richest coal fields mistakes have

been and continue to be made with implications for workers and shareholders.

Mining for lead was much more hit and miss and so it was just as well that the miners had another string to their bows of livelihood. What more natural than for the local farmers to turn their stout resolve and muscle power to a new job beneath their land?

Record has it that in 1550 Edward VI granted Robert and George Bowes the rights of mining within the Upper Dale and there are undoubted signs of mining activity dating back to this time. Thus new pressures were put upon the reserves of the local forest because wood was required to smelt the ore and it was then cheaper than coal hauled up with difficulty from the lowland collieries.

The practice of coppicing could well have increased at that time and coppiced woodland could have provided the necessary fuel without destroying the presence of the forest. The practice was already well ensconsed in the management of the English countryside, for in 1544 a statute proclaimed that twelve standard Oak trees must be left in every acre of forest put to coppice. This form of management thus opened up the forest canopy allowing the undergrowth, and especially Hazel, to benefit from increased light. The Hazel was then re-cut every ten to fifteen years leaving the stools from which vigorous new growth sprang up. The standard trees with their spreading crowns and shapely bent branches were used for the main timbers of ships, planks and wainscot. The Hazel wood was used not only for fuel but for a variety of other purposes: hurdles, fencing, wattles, the handles of a variety of agricultural and household instruments and indeed for anything for which stakes or bendable twigs were required.

Coppicing with its ten to fifteen year rotation has a marked effect upon the ground flora for the plants must be able to adapt from the deep shade condition of the old coppice to the much brighter sunlight, shaded only by the statutory twelve Oaks to the acre. From the Middle Ages Oak had also been coppiced to provide the structural components of timber-framed houses and farm buildings.

Well managed coppice is indeed one of those delights where it is almost true to say that Man has made improvements on the natural state. It is thus of interest

*Long Tailed Field Mouse with Lords and Ladies
and Ink Cap Fungus*

PLATE 54

to note that many of the plants which finally found refuge in the meadows would have themselves benefitted from the coppice cycle. Amongst these mention must be made of Wood Cranesbill, Hedge Woundwort, Herb Robert, Betony, Great Burnet, Melancholy Thistle, Globe Flower and Mealy Primrose; indeed the latter three may be found growing to this day in the lowlands in Hazel scrub. One also suspects that the practice of coppicing, or at least of cut and regrowth, had gone on ever since Man had brought polished stone into the Upper Dale.

With lead to be smelted and money to be made the interests so long invested in the continuity of mature forest could well have come to enclosure or at least to coppice compromise, as landowners and tenants alike turned their attention underground and to the business of producing food for the industrial towns. The end of the summer forest was in sight.

At first it was the local farmers who worked the mineral veins close to their enlarging fields, thus deriving double benefit from the same land, creating a new dual economy. However by 1758 the promise of riches underground brought eleven miners and their families from the then declining lead fields of Derbyshire to try their strong hands and expertise in opening up new mines within the Dale. Some found the climate too harsh and the going too tough and returned from whence they came. Others, like their leader Captain Wagstaffe, died in their new enterprise and some, namely the Ritfords later Redfearns, and the Staleys, stayed to make a living both from mining and the land.

The baptismal register again gives us a glimpse of these new colonists. July 1765, Robert of Samuel and Dorothy Retford; March 1766, Dorothy Ritford; March 1768, Peggy Ritford; February 1770, Lydia Redfearn; February 1771, George Redfearn; June 1773, Joshua. At first all their energies were spent on mining but by the 1780s they had taken lease of a plot of land adjacent to that on which the Robinsons had already lived for more than a hundred years. It was a hard life; lives were lost and no fortunes were made but the new dual economy worked as the others had done for almost 3,000 years. Hunter gatherer, hunter pastoralist, hunter farmer, and now farmer miner, for it would seem, and the same is true today, that the potential of the Upper Dale is insufficient to provide anyone with a living based on a single occupation. Perhaps this is the main reason why the area has always remained a backwater in human endeavour even though it is an outstandingly beautiful place in which to live.

With the coming of the Redfearns the last real wave of immigration into the Upper Dale began and we begin to pass from the span of written history into the span of human memory, facts passed down from generation to generation, by word of mouth. In with the tales of hardship, illness and death, which are the headlines of Everyman's existence, there are warm facts which tell of new ties linking families, neighbours, new societies within the Dale.

Whether fact or fiction (it doesn't matter) we are told that when the Redfearns arrived, strangers almost from another land, they found the house was clean and a fire had been set, forging a link of neighbourhood which eventually was sealed by marriage between the two families. When young George Redfearn (at no more than two years of age) disappeared from home the men from the mine downed tools to join in the search, fearing that he had wandered too close to the waterfall and had been lost. That particular story had a happy ending for he was found fast asleep within the chicken coop where he had been trapped earlier in the day. Accounts of the chores about the house, the joy of simple things at Christmas, the trials and tribulations of homebound scholarship fitted into a hard day's work, the ever looking forward to better things, backed with the inevitability of a life spent in the mine.

My favourite story, and I would like to think it true, concerns the eldest daughter of Samuel Redfearn who at the age of six bore the maturity of life around her frail shoulders. She could bake bread, sew, knit and even turn the heel of a stocking as well as any grown woman. She had in her wanderings through the scrub forest, which bordered their newly enclosed fields, found a family of Foxes and had befriended the cubs regarding them as puppy dogs. Hearing that the Duke of Cleveland's men were coming to destroy the lair she had gone and stood guard, defying the menaces of the dogs, the threats of the keepers and even money

Dorothea Redfern known as Lizzy

PLATE 55

offered if she went home like a good girl. The Duke himself came by and rewarded the little girl for her bravery with a crown piece and later with a green velvet dress and bonnet trimmed to match with pink rosebuds.

The Duke's men were doing no more than their job, keeping up the age old traditions of forest management, protecting next years fauns from their natural predators. Dorothy (called Lizbeth in the story) struck a blow for conservation, and blood money flowed as it would again to salve Man's conscience in his march forward to 'better' things. The story itself has a bitter sweet end and perhaps an added moral for conservation. Lizbeth's mother thought the dress too good except for extra-special occasions and so it stayed in its box almost unworn until out-grown.

Since reading the Redfearn saga, set down by one of the last of their descendants living in the Dale, I have been back to their farm and walked those acres for myself. It is all still there, field boundaries, wells; the forest has of course all gone but many of the flowers they saw still grow around the place. *'Great fat yellow dumplings, pale pink clusters of primroses and yellow ones also; cowslips, bluebells, lady's slippers and rock roses. There was hearts-ease, yellow, blue and white, and some with all these colours mixed up together such as I have never seen elsewhere. And there were orchids, one after another, beginning with great purple ones, spotted and speckled ones, pink and lilac ones, and last of all one in deep pink that gave off a lovely fragrance that filled all the house while they lasted. Father found others for us on his way to work, and in time we learned all their names.'*

The remains of the mine are still there beside the Upper Beck; in fact there is hardly a square kilometre of the Upper Dale that does not bear the mark of the new dual economy – field boundaries, old shafts, adits, drifts, spoil heaps and, perhaps most conspicuous of all, the hushes – a name which speaks of silence after years of feverish activity – that pockmark the living landscapes. A hush, and they come in all dimensions, consists of a Man made valley cut into the side of the Dale in order to reveal the bedrocks and, with luck, rich mineral veins. The method of their construction was to harness the erosive power of water by channelling it along the contours to a reservoir from which it was released downslope to do its best. The surge of water removed the peat top soil, the overburden of boulder clay and other glacial drift to reveal the bedrock loaded with promise. The size of some of these undertakings rivals even the overflow channels produced by the meltwaters of the Great Spring, cutting deep into the hard bedrock and suggesting that charges of gunpowder may have primed the way for the water to come.

Undoubtedly the first mines were located where mineral veins had been exposed by natural erosion. Hushing was thus no more than a natural follow on, and recent descriptions of hushes in Spain, which date from Roman times, show that the method has been in use for a long time with or without the help of gunpowder.

Mining, in whatever form, turned the clocks of soil formation back more than 10,000 years, providing large areas of well worked parent material ready for a new phase of pedogenesis. However, much of this new material added problems, especially when derived from lucky strike areas in which much lead had already been found. Lead is not a mineral which is essential to the growth of anything; in fact even in minimal concentration in the soil it can spell death to most forms of life. Because of this added problem in an already harsh climate some of these areas have remained bare of vegetation ever since, although today most of them do support some plants ranging from scattered clumps of the All Screwed Up Moss (*Tortella tortuosa*) to closed swards of stunted Red Fescue. This complex of more open communities has become the special home of four of the special plants of the Upper Dale, Spring Sandwort, Alpine Pennycress, Alpine Scurvy Grass, and Mountain Pansy. Whether it is just that they have evolved to grow in rough, tough places, or that their characteristic slow rate of growth means that in their life span they take up insufficient lead for it to become toxic, or simply the lack of competition with other ranker growing plants which allows them to thrive on this 'poisoned' land, we do not know. Suffice to say that they are there as part of the cosmetic surgery which evolution provides to cover up even the worst depredations of Man.

Lead Miner with Ponies

PLATE 56

Just as lead spoil heaps are unhealthy places on which to grow (at least for normal plants) so are lead mines unhealthy places in which to work, a fact which brings us to the next name in our story, one John Binks, but more of him in due course.

Mining took off in a really big way with the arrival of the London Lead Mining Company in the Upper Dale. They provided the capital, took much of the profit, and the locals, both old and new, did the work. The pattern of life of the majority of the miners was much the same as that of the Redfearns; a small farm run by the family which provided, at least, a basic subsistence diet, plus wages from the mine which held out high hope for a brighter future, a lucky strike always just around the corner!

The majority of the new mines were a long way from habitation, way up in the outbyeland and so the practice of *walleting* grew up. Each miner would go to live at the mine shop, carrying with him enough food and provisions for a hard week's work. These were packed into the ends of a long cloth sleeve or wallet which was carried slung across the shoulders. They returned home, wallets empty, only at the week-ends to go to church and do the heavier jobs around the farm. The lead ore was brought back down to the smelt mills in panniers slung across the backs of hardy hill ponies, descendants of those which had worked on the building of the Roman Wall. Life at the shops was hard, especially in the winter, but there was that certain camaraderie of purpose which appears to develop only amongst men working underground. It may still be savoured even in the most modern mechanised pit, and *bait* as they call it, is not only a time for eating but a time when the social hormone which binds every hard working society together flows freely.

It was out of this sense of relationship in hard work that all that is good in the Trade Union movement evolved and nowhere more so than in the coal mines where the workers had direct contact with the energy stores of the past, energy flows of the present, the basis of a possibly brighter future.

Sitting in the crumbling affluence of a fossil fuel based economy it is difficult to understand why these people put themselves and their children to the indignities of life underground. Surely a life based on the produce of fresh air and sunshine was much better. The problem is that within the limitations of such a subsistence there is no further view, no hope of better things on the horizon, except perhaps that a prince would come and carry you away if you happened to be a girl. The hope of a good growing season was always there but without the stability of the forest and the excitement of the hunt, there was nothing to which to turn in the bad winters. The mine filled both these gaps.

Likewise, as the population grew there was insufficient land from which to supply even subsistence needs and imported grain began to play a larger and larger part in feeding the workers.

The drug of this new dual existence was energy stored up by forests of the past and the fact that this energy could now be exchanged for money, food, and other things. No, that is too simplistic a review. Fossil fuels were no more than the vehicle which allowed this new development: they were not the driving force. Survival dragged people from the inadequacies of the land to the hopes of factory towns built round the mines, but that was not all. The message of The Great Seasons has again and again been, not one of survival, of hanging on to what you have, but of new enterprise, of reaching out to better things; not the dogma of survival of the fittest but the true philosophy of evolution that *wherever there is potential it must be used*. The potential was there, dragging evolution on from the first living chemicals up to the consciousness of *Homo sapiens*, out of the sea onto the land, from the pioneering communities of the Great Spring up to the affluence of the summer forests, each successive stage, or phase, having the ability to flow more energy, exploiting the life potential to the full.

Man, as part of the evolving system, could not kick the habit; he could not disobey the rule. Brain power and dextrous hands gave access to a new and seemingly inexhaustable supply of energy; consciousness spicing the whole endeavour with the expectancy of that lucky strike and more-to-come just around the corner.

The coalfields of North East England had first been opened up in earnest to

Little Owl with Emperor Moth in old Lead Workings

PLATE 57

supply the energy requirements of Londoners when they ran out of local coppiced woods in the early 1600s. This had provided new opportunities in the coalfields and people came to work there and raise their families. The Upper Dale responded with meat, hides for leather and wool for clothing. The final destruction of the forest began. The mineral veins then poured out their riches, increasing the income of the London based shareholders and the numbers of the local population.

It was not all bad, and in this 'you gets your money and takes your chance' way of life there were both winners and losers. The Lead Company put a certain amount of revenue back into the Upper Dale: transport, schooling, pastoral and medical care were all improved. The new roads which cost at least £12,000 linked up the scattered communities ensuring that both the vicar and the doctor could make the journey in the comfort of a trap and the lead could be exported by cart. This also made the import of coal and other goods much easier, strengthening both the import and export economy, which increased the need for cash and opened up new opportunities of employment, so much so that the population census of 1851 found the Upper Dale with as many people as it had ever supported, nine hundred and four to be exact, which makes just under five persons to the square kilometre.

At this time sixty eight families made up the human genetic stock of the Upper Dale and it is of great interest to analyse the population statistics to attempt to ascertain which are the oldest residents. On the basis of abundance and widespread distribution the following families would appear to top the list. Behind each family name the number of members of the family living in the three main centres of population as you pass up the Dale is given.

Dowson – 15, 0, 49; Tallentire – 27, 7, 19; Anderson – 24, 0, 27; Bainbridge – 23, 21, 3; Scott – 26, 17, 0; Watson – 0, 0, 42; Walton – 11, 31, 0; Teward – 12, 10, 18; Beadle – 29, 5, 0; Bell – 15, 16, 0; Hutchinson – 19, 1, 9; Tarn – 22, 6, 1; Allinson – 16, 10, 2; Rumeney – 0, 0, 25; Collinson – 3, 12, 8; Gargett – 7, 16, 0; Robinson – 0, 13, 9; Horn – 11, 0, 13; Howard – 0, 14, 5; Cousin – 0, 0, 19; Brumswell – 17, 0,

0. The first parish records show that all these families save three were in residence in the Upper Dale one hundred years before.

However when the list of occupations is examined it is found that only four are full time farmers, six agricultural labourers, three gamekeepers and one gamewatcher, these together supporting fifty one people. Compare this to two hundred and seventy one directly employed in lead mining which thus supported the bulk of the population and the real picture begins to emerge. Not of a population supported by the productivity of a certain area of land but one dependent on a volume of resource. As few if any of the mines penetrated deeper than 50 metres the volume of that new resource space may be calculated at a gross 10 billion cubic metres. In itself a meaningless figure, for much of that volume contains no useful minerals of any industrial importance, in the same way that much of the land surface was of little or no use for agricultural production. It does however emphasise the fact that during this phase of his development Man was beginning to be a major geomorphological force reshaping sections of the landscape and concentrating certain elements in modified form in certain areas.

The use of 'craw coal' won from the meagre outcrops of the Upper Dale must have released sulphur dioxide into the air. Now with coal more easily imported it was mixed with peat stripped from the hills to smelt the ore. The increasing pollution of the atmosphere killed off many of the lichens around the smelting shops. Coal was also used to produce lime for the revitalisation of the meadowlands against the increasing loss by export and to leaching. Limekilns thus became almost as much a feature of the enclosed lands as spoil heaps and abandoned mines were of the upper fells. This period also saw the start of drainage and other bottomland improvements which began to add a certain symmetrical beauty to some of the fields. The hollows caused by the laying of the drains are often picked out by King Cups in the spring and by other members of the Buttercup family later in the year, and the rocks removed from their depths went to build the dry stone walls. All these were changes to help support the human population.

The fact that amongst the nine hundred and four inhabitants only two were

Hedgehog with Lapwing's Eggs
    Strata exposed by quarrying and the Whin Sill in the background

PLATE 58

grocers and both of them had other jobs, one a carrier, the other (what else?) a miner, certainly lends weight to the supposition that most of the 119 habitable dwellings within the Upper Dale had access to a plot of land on which to grow at least some of their food.

The other trades and professions supported by this immense volume of underground activity were in descending order of number rather than importance. Domestic servants, 22; stonemasons, 5; servant labourers, 4; dressmakers, 4; wallers, 3; herds, 3; innkeepers, 2; blacksmiths, 2; masons' apprentices, 2; and one of each of schoolmaster, curate, ironworker, tailor, parish labourer, hind, highway labourer, farm servant, cart driver, journeyman (paid by the day), shoemaker and apprentice, ironstone labourer and last, but by no means least, a slate pencil maker and his labourer.

The work of these latter two souls, by the name of Tarn and Wilson (the labourer who was 11 years of age lived in the farm nearest the quarry) provided slate pencils enough for 105 scholars. They attended one of two schools then open in the Dale. The lower (in altitude) of these had a part-time schoolmaster/farmer by the name of Robinson, the other had a full-time master by the name of Jones, from Wales, one of only six foreigners living in the Upper Dale. The others were four labourers from Ireland and a curate from Scotland who enjoyed the living of St James and St Jude which was all of £254 per annum.

The enclave of learning links, through its green slate pencils, the endeavours of the future with the true foundation of the past laid down more than four hundred million years before. The expansion of knowledge and especially of the three Rs made the publication of a Dale newspaper both practical and necessary. So the affluence of lead spawned the quicksilver columns of *The Mercury*, a tabloid which has set down the happenings of the Upper Dale against the background of world change ever since its first edition on June 1st, 1854.

However, long before the first edition of the Dale's own local newspaper came off the press, many facts concerning life in the Upper Dale had been set down in much more wordy publications.

John Binks and the Days of Discovery

In the years between 1670 and 1863 much more than lead had been discovered in the area. *A Botanist's Guide to the Counties of Northumberland and Durham* written by three eminent local naturalists, Winch, Thompson and Waugh, and published in 1805, lists many of the special plants now known to grow within the Upper Dale. Articles in many of the leading botanical journals of the day include numerous accounts of visits to the area by leading botanists, and the opening of numerous accounts of visits to the area by leading field botanists, and the opening of a branch of one of the first passenger railway lines in the world made access to the area that much easier.

My favourite account is of two days botanising led by the two James Backhouse (father and son) which took in all the then known treasures of the Upper Dale and discovered more to swell the lists, including the Bog Sandwort which today ranks amongst the rarest plants of Britain.

One of the first things I did when I came to live in the area was to make a plant pilgrimage retracing the steps of their marathon of discovery. The first day was a very long 43 kilometres, part lost in cloud, but well worth every minute and I could report that all is well; the plants are still there growing in abundance. The second day was even better for it took me through the richest area of the Upper Dale and the sun was shining into the bargain. The following list comes from my own field notebook.

The special plants of the Upper Dale seen on a 'pilgrimage' made over the last two days of August one hundred and twenty years after the epic journey of the Backhouses in 1843. I retraced their route looking for the plants in the order in which they found them.

Cloud Berry, Northern Bedstraw, Alpine Penny Cress, Bird Cherry, Globe Flower, Variegated Horsetail, Shrubby Cinquefoil, Melancholy Thistle, Hair Sedge, Broad Leaved Cotton Grass, Alpine Bartsia, Scottish Ashpodel, Alpine Bistort, Spring Gentian, Small White Orchid, Yellow Marsh Saxifrage, Parsley Fern, Bear Berry, Mountain Avens, Three Flowered Rush, Hoary Whitlow Grass, Hoary Rock Rose, Alpine Club Moss, Fir Club Moss, Starry Saxifrage, Hairy Stonecrop, Alpine Scurvy Grass, Chickweed Willow Herb, False Sedge, Bird's Eye Primrose, Blue Sesleria, Alpine Meadow Rue, Mountain Everlasting, Holly Fern, Brittle Bladder Fern, Green and Black Veined Spleenworts, Yellow Mountain Saxifrage, Rose Root, Alpine Cinquefoil.

I looked in vain for *Woodsia ilvensis*, Jacob's Ladder, Snow Saxifrage, *Cytopteris regia*

Sugar Limestone

ing Gentian

John Binks

False Sedge

Little Bog Orchid

Roseroot

Alpine Bartsia

Common Juniper

Bog Whortleberry

Hoary Rockrose

Cowberry

PLATE 59

and Horseshoe Vetch. I could only find one small plant of Holly Fern well grazed by the sheep. I must confess that I have never mastered the complexities of the Hawk Weeds and cannot therefore confirm the earlier records. In my journey I came across a number of the special plants which have been recorded for the Upper Dale since 1843: a list of these is appended in the order in which I found them and wherever possible the date of the first record and the name of the recorder is given.

Meadow Horsetail (1844, J. Backhouse)
Mountain Pansy
Alchemilla wichurea (1947, S. M. Walters)
Alchemilla glomerulans (1947, S. M. Walters)
Alchemilla acutiloba (1922, A. J. Willimott)
Alpine Foxtail (1959, A. Eddy)
Alpine Willow Herb (1862, J. Backhouse)
Alpine Forget-me-not (1862, J. Backhouse)
Tall Bog Sedge
Heath Violet (1862, J. Backhouse)
Rare Spring Sedge (1949, T. G. Tutin)
Sea Plantain
Sea Thrift
Bitter Milkwort
Alpine Rush (1903, G. C. Druce)
Bog Sandwort (1844, J. Backhouse)
Alchemilla acutiloba (1933, A. J. Willimott)

All those rare plants in just two days and 86 kilometres of walking! No wonder botanists have for the past 120 years regarded the Upper Dale as a very special place.

James Backhouse junior wrote in *The Naturalist* for 1844 concerning the botany of the Upper Dale, telling how many of the plants were first discovered not by himself and his father but by one John Binks, a miner who supplemented his income and his unhealthful occupation in the mine by spending one day in every five at large upon the fells collecting plants. His main interest lay in plants like Rose Root, Juniper, Bear Berry, and the like, which were of medicinal value. He passed the *materia medica* on to the local apothecary and knowledge of his other finds to a local clergyman, probably the Rev J. Harriman, and to James Backhouse senior, whom he had led on 'many a botanical ramble'. The plants, especially the more showy ones, must have been well known to the locals, at least by sight if not by name, and at one time there was almost a tradition in the houses as to who could put on the best show of Gentians. The practice was to cover a ball of clay or moss with Gentian flowers and set it on a crock in the window for all to see. When this practice started and died out I do not know and though as a conservationist it makes me shudder I cannot help but wish that I lived back in those days when the human population was small enough to allow all who would to pick a bunch of wild flowers.

It wasn't only in the Upper Dale that botany became so popular. Botany was a growth interest and 1828 saw the publication of the sixteenth *Flora of the British Isles* and some of them had run to several editions. These included among their number a part work of 36 volumes with 2,592 coloured plates drawn by James Sowerby with a text written by Sir Jas. E. Smith. The perfection of the book, the *de luxe* volumes bound in hand tooled green leather, is to me the epitome of the products of the gilt-edged life of those times. It seems like sacrilege to say that it was just one of the many books of ready reference for those who could read either Latin or English and could afford to buy them.

Certainly there must have been spare cash about at that time and not just amongst the very wealthy, for on March 12th, 1871 *The Mercury* reported that many holiday makers passed into the Upper Dale *en route* to the two great waterfalls.

Since those halcyon years when botany developed to be not only a gentle person's pastime but the mother subject, the cornerstone of agriculture, horticulture, plant breeding and the plant sciences, new discoveries have been made in the Upper Dale with surprising regularity and I am sure there are more to come.

Of all these the one I still find most fascinating is the discovery by Tom Hutchinson (now an eminent Professor of Botany in Canada) of the Dwarf Birch, growing on the same spot where pollen evidence proved that it had flowered for the past 10,000 years. The reason for my surprise, and I must admit chagrin, stems from the fact that on my pilgrimage visit I had walked right across the area in which it was later found growing. I had been on my way to see the two rare Bog Sedges and *Haplodon wormskjoldii* and my feet must have passed within metres of those five brave trees, descendants of the genetic stock which had produced the first mini forests of the Dale.

Like the glaciers of the Great Winter lead mining came, reshaped the Upper

Traditional Gentian Bowl

PLATE 60

Dale, and went. Its peak of destructive affluence was in the 1850s and 60s, after which it dwindled away and with it the population itself went into decline. In 1894 Whellan's directory records the population of the Upper Dale to be 675 and the fact that, except for the Rumineys and Waltons, all the top twenty families were still in residence. Perhaps most significant of all it states 'The population of the Parish is chiefly employed on the land as well as doing a little in the lead mines.' The only other occupations listed are one vicar, two schoolmasters, and two victuallers, each running an hotel, one of which boasted a sporting establishment.

Whether plant hunting came under the term of sporting pastime, I do not know, but it certainly waxed in popularity because in editions of *The Mercury* of 1880 there are the following two statements; *12th May, Mr Scarth (on behalf of the Duke of Cleveland) issued a notice prohibiting the picking of plants on the estate; 25th August, the flora of the area has been depleted since the opening of the rail links.* It is also worth noting that on October 13th in that same year there was a report of a rumour that the local Water Board wanted to construct a balancing reservoir in the Upper Dale – shades of things to come!

The best skating in living memory was also seen in 1880 when a Dr Mitchell and Mr Richardson skated for over thirty kilometres along the river. The columns of the paper also discusssed that in 1855 the ice on the river had been so thick that a heavily laden cart could cross in safety and that in 1740 horse races had been held upon the river with ten entries for three heats (or should they have been called colds) each over two miles while the great waterfall had a hollow icicle over twenty yards in circumference.

If the winter of 1880 was bad (or good) depending on your point of view, the summer brought a drought which was also the worst in years and had a catastrophic effect on the Salmon fisheries. It was not only the weather of the Upper Dale which remained changeable, for in that same year, February 4th to be exact, *The Mercury* reports further changes in the life style of the area. The past twenty five years had evidently witnessed a marked improvement of agricultural practice, especially extensive draining on the Duke of Cleveland's land. During this time the cultivation of cereals and turnips had become unprofitable due to the cost of horses, high wages and a succession of bad seasons; in consequence the land was all put down to permanent pasture. From this point on it would seem that even the bulk of the locals would now be dependent upon 'imports' for much of their food.

The old copies of the newspaper make fascinating reading for they trace the local happenings in detail and show how both the weather and Man's utilisation of the resources of the Upper Dale were in a state of constant change.

The final transposition of extensive tracts of forest into enclosed farming land must have turned the latent hunting instincts of both locals and visitors alike up towards the potential of the high fells and to one game bird in particular; its name, *Lagopus scoticus*, the Red Grouse. It lives both amongst, and predominantly on, Heather, especially on the drier sections of the blanket peat. In part it depends, as do many other members of the community, on the growth cycle of the dominant Heather, for it requires both old and leggy Heather in which to nest and shelter its young from the strong winds, and young succulent Heather shoots on which to feed. Realisation of these facts allowed Man to begin management of the resource to his own ends. This included not only the removal of as many would-be predators as possible, but also the regularisation of the Heather by controlled burning. The aim of this practice is to turn the Grouse moor into a patchwork of small stands of Heather, each one of a different age so that each landscape unit provides as much potential for the Grouse as possible. It is difficult to say what the original carrying capacity of the natural moorland was; a figure of one adult Grouse per hectare would probably be an overestimate. However, in a well managed moor the number could well go up by an order of magnitude, especially in a good year. The population of Grouse, be it natural or man managed, appears to be, in the main, regulated by resource rather than predation. Each autumn the male Grouse will 'stamp' out the bounds of his territory which he then guards jealously from all except his chosen mate. Any Grouse which does not manage to lay claim either to a territory or a mate dies of starvation during the winter, thus effectively controlling the population. Most surprising of all, the size of the

Cross leaved heath
and feather moss

Red Grouse

Spaniel

Grouse Chick
Pheasant Chick

territories and hence the carrying capacity of the moor alters from year to year, and there is good correlation between this and the success of the *next* season's production of Heather. Much research has gone into trying to understand how the birds know what the Heather holds in store for the next year, but so far it has not penetrated their method of long distance forecasting. *Lagopus scoticus* is thus the very model of resource planning and one of which Man should take note.

I learned most of what I know about the ecology of the Grouse moors by listening to one of the great keepers of this century, Walter Curry. It was fascinating to tread the moors with him and learn their secrets. 'Conditions are only exactly right for burning on a few days each year, not too dry, not too wet, with sufficient wind to fan the flames and keep them on the move but not enough to send them racing across the moor. The fire once lit must be contained so that the burn regeneration patches are not too large.' He would bend down and feel the Heather, point to a stand burnt three years before that was uniform like a bowling green and murmur content of a job well done. He tolerated our attempts to get a scientific measure of the standing crop and productivity of the Heather but I am sure that, like the birds, he knew all the answers.

With increased mechanisation and higher wages upland management of the right sort, which is very labour intensive, is a thing of the past. Though August 12th may still dawn glorious with the almost hallowed whisper 'the gentlemen are here' running through the Upper Dale and the Grouse bags recorded in *The Mercury*, I don't think we shall ever see another set of seasons like those of the 1930s when climate, Man, Heather and *Lagopus scoticus* got together and filled the game bags and record books.

Calluna vulgaris, Lagopus scoticus, and *Homo sapiens* together have maximised this unit of production, changing the course of evolution of certain populations of all three. The plant to greater tolerance of regular burning, the birds to lie and fly as low as possible when the beaters move across the moor and Man to be swifter in reflex and more precise in the manufacture of his armoury.

I have always wanted to own twin Purdeys, hand crafted for my use. They are as perfect a product of Man's industry as any other, an extension of our muscle and nervous power, a perfect foil, not of defence or attack, but of the livelihood of hunting which has its roots deep in our ultra ancestry. Unless my family were starving and there was no other way of obtaining the necessary food I would not use them on another living thing and would content myself with skeet pigeons made of clay.

Please don't get me wrong: I am not glorifying the hunt but I am not condemning it out of hand. If I were a vegetarian perhaps I could, with my hand on my heart, say, no it shouldn't happen, it is wrong to enjoy killing animals. I am not and I enjoy the succulent spring lamb culled from those same hills, so who am I to decry another compatible form of management which keeps the harshest area of this harsh backwater a viable unit within modern society. I would remind the vegetarian that Man is by descent an omnivore and a decision to abstain from eating meat is no more nor less part of Man's free will than another's decision to train as a marksman and gain enjoyment from the excitement of the hunt. The only rational argument for the one and against the other is that the hunting instinct reaches back towards a more affluent past while vegetarianism looks forward to a future in which the resources of the world will be so strained that we must all perforce eat at the lowest level of the pyramid of energy.

I cannot help but feel that Man has a lot to learn from the Red Grouse.

Resources in Conflict

Modern management of the Grouse moors includes the use of gigantic ploughs which cut deeply into the peat, sapping at its life blood. The purpose behind the practice is partly to drain the moors thereby increasing the productivity of the Heather and over much of the Upper Dale the drains can be seen like black and silver serpents snaking their way across the contours.

They could well be a strategy for future catastrophe, for over much of the Pennines the blanket of peat is already in massive recession and large tracts of once productive moor are brown wastes of peat hags, a wet desert of a little use to Man

Black Grouse with Chicks and Eggs, Mountain Pansy, Northern
  Eggar Moth and Yellow Mountain Saxifrage

PLATE 62

or nature. In places the remains of the past affluence of the summer forests have been eroded from the bottom of the peat. Their rootstocks lie spreadeagled on the jumble of boulder clay and other glacial deposits, reminding us of what has been and what is still to come.

A living peat blanket holds water in store against the force of gravity: it is a reservoir moulded to fit the upper contours of the hills and covers all except the main drainage axes and major breaks of slope. It must therefore act as a balancing reservoir smoothing out the fluctuations in the water flow of the streams and of the river itself. The effect is however minimal because if the peat is to continue growing it must remain saturated throughout the year and so the reservoir remains almost full, the gates, in essence, shut. In wet weather flash run-offs still occur, swelling the great waterfalls to magnificent proportions; the 'freshet' rushing down the river scours its course, tearing at the developing vegetation and keeping the banks open. During drought the living reservoir holds onto its own, losing some water to evaporation and some to drainage. The latter boosts the river flow but marginally until it rains again. A sponge when full of water can hold no more and the majority is only released when the sponge is squeezed.

Partial drainage of the peat blanket must improve its reservoir capacity, for it provides phreatic freeboard which is filled by rain, and spillways for the same water to flow towards the river during drought. The problem is that drainage (or 'gripping' as the locals call it) opens up the surface layers of peat to oxidative decay and the end is then in sight.

Mismanagement of the Upper Fells for short term gain has led to the almost complete eradication of two plants which in places make up the bulk of the peat. They are *Sphagnum imbricatum* and *Sphagnum fuscum* and as both grow into large hummocks which protrude above the general water table they are very susceptible to fire. With their living presence gone, gripping and further fire can only lead to breakdown and erosion of the peat, the end product being a wet desert with no water holding capacity, of little use to Man or nature.

In all this talk of management, the river itself must not be forgotten for it is the main route by which the majority of natural exports leave the Upper Dale. The final products of erosion, be they animal, vegetable or mineral in immediate origin, have always found their way into this ebullient water course. Like all rivers which gather their water from a mountainous catchment its pulse can be very erratic and flows vary between over 40,000 million litres per day to less than 100 million litres per day, measured just above the point to which high tides bring salt water up into the estuary. Such erratic behaviour must affect all life within the river's compass, an important part of which are the anadromous fish, Salmon, and Sea Trout, which make their annual way upstream to spawn.

These large fish must have always offered a ready supply of fresh protein up to the foot of the great waterfall. Salmon used to be so plentiful that they were regarded as the poor man's staple protein and had been fed as 'red flour' both to Pigs and Hens. Again and again the columns of *The Mercury* lay blame on the construction of a dam lower down the river for ruining the fishing in the upper reaches. This may well have been a factor in the slow decline of these migrating fish but the real blame must be laid at the feet of increasing population and pollution of the river, especially of its lower reaches where it emerges to the sea through an area of heavy industry.

Fish can 'breathe' underwater, their gills drawing on the meagre supplies of oxygen present in solution. If the dissolved oxygen is used up then the fish will 'drown'. Organic matter, in whatever form (leaf litter, the carcasses of animals, animal or human excrement), is potential for the decomposer side of life. If such material is voided into a river, that potential will be used as the organic matter is broken down by bacteria and other organisms which will multiply in the water body. All this extra (decomposer) life increases the normal demand for oxygen within the water body. If the stores are used up quicker than they are replenished by solution diffusion and photosynthesis, the water will become deoxygenated and that section of the river will become a barrier to migratory fish.

So it was that a survey carried out in the estuary between 1929 and 1933 showed 'that it was heavily polluted with sewage, so much so that the central

Lapwing with Chicks

PLATE 63

section of the estuary was partially deoxygenated'. Added to that much of the estuary was also found to be toxic to fish life. The evidence pointed to cyanide (discharged from coke and gas ovens) as being the main cause of fish mortality.

Although the main effects of the Industrial Revolution had come and gone in the Upper Dale, industry had continued to expand and diversify near the mouth of the river but it was not until 1967 that the rumours mentioned in *The Mercury* of 1880 became the reality of an Act of Parliament which condoned the construction or a balancing reservoir in the Upper Dale. Its main purpose was not to cleanse the estuary but to improve the yield of the river during periods of low flow so that sufficient water could be abstracted to quench the thirst of an ever expanding industry near the coast. Once again the resources of the Upper Dale were to be put to use with little or no hope of direct benefit to the local population.

The Parliamentary Bill was not made law without a fight and central to the arguments against the exact location of the dam and its impounded water was the uniqueness of the flora of the area. It would be pointless to drag up all the details save to say that they were argued by some of the best lawyers in the land. The case for the reservoir revolved around the 'fact' that the economy on which the viability of Britain depends demands that chemical and other industries must expand. To be able to do this they need water and the most sensible and economic place to build the necessary reservoir was at a central location in the Upper Dale. The arguments against simply stated that the botany of the Upper Dale was unique; it was an area of outstanding natural beauty and interest, a living workshop of evolution, of immense value to education, research, and amenity, a very special part of Man's natural heritage.

Experts were called in and expertise was conjured up. All applied their own statistics to the same data and each came up with remarkably different results and recommendations. Newspaper headlines got bigger and bigger. Tempers were frayed, reputations were made and marred; the facts were weighed in the scales of public enquiry and a decision was made to build the dam. Part of the resource of the Upper Dale which had to date flash flowed to waste, polluted by the salt waters of the sea, was to be impounded and put to use in the service of industry.

In a world which was taking heed of *Silent Spring,* shouting Serengeti shall not die, supporting the newly founded World Wildlife Fund and in a country which boasted a government-funded Nature Conservancy, it was tragedy indeed. In respect to these facts of nature, the main beneficiaries of the water which would one day flow from the reservoir gave a sum of £100,000 to be held in trust to finance ten years of research into the natural history of the area.

Ten thousand pounds a year and only five years to go before the dam would be finished, the reservoir full and the main, irreparable damage done. What a challenge!

This is not the place to detail that research nor to mention all the people who responded to the challenge and the new potential. Their work has and still is being put on record in the leading scientific journals. However, it does seem appropriate at this juncture to review our understanding of the hows and whys of the natural history of the area, which was greatly enhanced over this period of hectic work.

The origin, the how this unique flora came to be, has unfolded with the story of the Four Great Seasons, a story of climate, geology and landscape history. The exacting work of the pollen people led by Judith Turner added the meat to the bare bones of theory and provided us with as detailed a picture of the evolution of a living landscape as any we have to date.

Our understanding of the *status quo* has its foundation in a classic paper by Professor Donald Pigott who above all others has upheld the traditions of British ecology founded by Sir Arthur Tansley and Professor Willy H. Pearsall. It set down a basic description of the vegetation, describing in detail both the pattern of and the process by which the diversity is maintained. He detailed the complex of ecological conditions, acid, alkaline, lime rich, lime poor, wet, dry, heavily grazed, ungrazed, man managed, mismanaged, semi natural, facing north, south, east or west, and all points, permutations and combinations in between: a whole complex of units and their boundaries, variety both in space and time in which a whole society of plants can coexist.

Badger with Woodcock Eggs and Violet Ground Beetle

PLATE 64

Likewise the vision of David Valentine's school of experimental taxonomy had shown the Upper Dale to be an island in which the rare species live in glorious isolation, genetically out of contact with other populations. Thus it is that any special traits which these populations inherited from their pioneering ancestors could have maintained or developed undiluted by the influence of genetic information brought from outside. In essence, the special species had, in isolation, each gone along its own sweet way of evolution in its island home. It is perhaps sufficient to say that many of the population do show variation, for remember this is the stuff on which natural selection acts. There are differences which set the plants of the Upper Dale apart from those of other localities across the world but none are sufficient to warrant the erection of a new taxon even of sub-specific rank. The evidence is however there that evolution is at work within the Upper Dale and that this island refuge contains a unique genetic resource for the future.

The fact that the majority of these special species of upland plants can be grown in the lowlands in a well ordered alpine rockery indicates that they are not intolerant of warmth and long growing periods. However, any gardener will tell you that the most important part of the management of any garden in which exotics are to grow is a regular programme of weeding to keep the more natural denizens of the area at bay.

Gordon Manley had shown that the climate of the high western ridge of the Upper Dale is marginally sub-arctic. His description is a very apt one: 'We therefore form the conception of an excessively windy and pervasively wet autumn, a very variable and stormy winter with long spells of snow cover, high humidity and extremely bitter wind alternating with brief periods of rain and thaw. April has a mean temperature little above freezing and sunny days in May are offset by cold polar air; while the short and cloudy summer is not quite warm enough to support the growth of trees. Throughout the year indeed the summits are frequently covered in cloud.' He goes on to say that 'a relatively slight rise in the frequency of warm anticyclonic summer weather would allow a rise in mean temperature sufficient to support the growth of trees'. We know it has happened in the past and that the high western ridge was covered with Elm, and Oak. So why is it today that over 300 metres lower in both north and south facing slopes, ecosytems harbouring arctic alpine plants are widespread in conditions which might therefore be regarded as marginal to their survival? It must also be remembered that those same communities also find room for a number of plants from the warmer south.

What then is responsible for the maintenance of these special gardens of the Upper Dale?

Open tracts of land are common occurrences along flow lines and around springs and seepage lines and there the soil is open to frost action every winter. The potential of such places is there each year to be recolonised by descendants of the plants which originally filled the Upper Dale early in the Great Spring. Similar open areas have been created more recently by mining and in places the soil is still in the process of recolonisation and so presents an environment of changing opportunity for plants intolerant of competition. The open character of all such sites is aided each year by grazing pressure which not only removes a proportion of the annual production but tramples the soil into the bargain.

At the other end of the scale are the meadows managed by a tradition which demands neither too much, nor too little from the land and so maintains stable communities of great beauty.

The really surprising thing about the vegetation is the closed swards of grassland which contain the real wealth of the Upper Dale's flora. They are unique and that uniqueness depends as much on what they don't as on what they do contain.

As the reservoir was going to destroy some of these very special communities, a certain amount of destructive research could be contemplated. Study showed that the abundance of rare plants is inversely related to the standing crop of the vegetation. The greater the standing crop and productivity of the grassland sward the fewer rare plants did it contain. Analysis of the floristic makeup revealed that the least productive swards were the poorest in the ranker growing grasses which

Red Fox with Water Vole. In the background the Sugar Limestone

PLATE 65

typify the lowlands, being dominated by Blue Seyleria and a whole variety of sedges. What then holds back the production and keeps the more lowland dominants at bay?

Dr Jeffries, working with Piggot out of nearby Lancaster, was able to show that the presence of lead in the soils was of key importance. Analysis of the soils and the plants growing on them showed that although there was ample phosphate available within the system the plants were unable to take up sufficient to maximise their growth. Everything pointed to the conclusion that the high levels of lead within the soils, especially those derived from the sugar and other forms of metamorphosed limestone, adversely affects the uptake of phosphate by the plants. Feeding experiments in which fenced plots were fertilised with phosphate backed up the theory with some startling results. Red Fescue took over dominance in many of the plots, increasing the standing crop, making life for the rare plants very difficult.

The mineralising fluids of 295 million years ago together with the intrusion of the Great Whin Sill had done their job well, providing tainted ground on which only a limited variety of plants could thrive.

This was not the end of the story for other experiments carried out by John Waughman and Evangelos Kookorinis from the University of Durham showed that climate also played an important role. They erected mini-greenhouses over a range of the communities developed in their most marginal location, south facing slopes around the 460 metre contour. Within the greenhouses production and standing crop were both significantly higher; the plants showed an increased uptake of phosphate even though the soils were not fertilised, and what is more the rare plants began to disappear, competed out of existence by the ranker growing grasses.

Thus a complex interaction was revealed; both the climate and much of the soil of the area presented problems for healthy growth, especially of lowland plants, the one exacerbating the other. If one problem is alleviated the plants can overcome the other, but in combination the problems are too much; the communities remain open in character and low in productivity and full of interest.

As the detailed ecological work went on, the hectic business of recording the detail of the complex vegetation that would be lost continued in earnest. This meant that every day was precious and work continued. Whatever the weather, one and often two lone figures were there each draped in a variety of waterproof attire which included non-absorbent knee pads to speed and cushion them on their way. Their names, Margaret E. Bradshaw and Allison Jones. They were the experts who spearheaded the attack backed up by gallant bands of recorders who worked throughout the growing season. They crawled and counted, marking each rare find with an appropriately coloured stick; these markers then formed the basic data for the construction of maps and the erection of a classification of the vegetation, identifying what was about to be lost. There is no time to mention all the people who gave up their time and expertise in the service of the Upper Dale nor to mention all the facets of the research. Suffice it to say that all that was humanly possible was done in the time available.

I must however make special mention of one person without whose forthright presence much of the survey work would never have been contemplated, let alone carried out. Margaret Elizabeth Bradshaw is a plant taxonomist by training, a teacher by profession and a conservationist by dedication. Her special interest was appropriately the Lady's Mantles. Of Yorkshire farming stock, the air of the open moors flows in her veins and with it, detailed appreciation of those special relationships between crops and their environment. Above all she passionately believed that the natural resources of the Upper Dale were something very special and must be conserved at all costs. The decision to build the dam simply strengthened her belief and so the cause gained a real and worthy champion who would have chained herself to any railing and did in fact defy the might of a bulldozer as it advanced towards one of her research areas. Like Elizabeth Redfearn she stood her ground and gained her just reward, not a dress of green velvet trimmed with pink rosebuds but the knowledge that no shred of information concerning the patchwork of green, bejewelled with the pink of Mealy Primrose, the blue of

Mealy Primrose with a Pigmy Shrew

PLATE 66

Gentian and all the other tints and hues of that pastiche of history, were lost to the annals of botany.

I made my last visit to walk along the wheel of water that would be the first to disappear. This famous landmark is, or rather was, the only real meander in the river's upper course, a mark of its first freedom to flow uncontained by steep fells across the series of alluvial flats from which it eventually tumbled over one of the waterfalls. This section of the valley had been reshaped first by the glacier which had filled the old course of the river with drift and sent it falling precipitously over bare bedrock, then by Man, the local mines within the basin adding to the alluvial deposits. I walked up-valley so that the Great Dam was behind me and reflected on the wheel which itself reflected the moods of the weather and the time of day by turning from silver through gold to darkest brown. I thought too of another great character of the Upper Dale, Hugh Proctor, contemporary incumbent of the local parish and an eminent botanist in his own right. The wheel was his favourite haunt and he had worked its waters and its banks, listing the aquatic vegetation which took advantage of the still, shallow water. He and other workers listed in all 26 species of macrophyte (large aquatic plants) growing within the aegis of the river. These included one rare plant almost on its southern limit, the Northern Water Sedge, only discovered in the area in 1969. All this would soon be lost, the question being could the plants 'rise' to the occasion and recolonise the margins of the reservoir and would the projected increase in visiting water birds (which had been cited as a reason in favour of the reservoir) bring in these and other species once again? I allowed myself the luxury of collecting one flower of Marsh Cinquefoil and wondered whether in its myriad cells there was new genetic information ready to be put to the test of natural selection.

Death brings new life, the rule remains unchanged, and in its applications knows no bounds. It's there, a cornerstone of life throughout the Four Great Seasons. So when in June 1970 the White Dam was completed, the death/life process was set in motion once again on a massive scale. The irony of water, giver of all life, now turned slow harbinger of death, crept up across the contours. 'GO

BACK, GO BACK,' the Red Grouse cried but all to no avail; they lost their rights of territory as the brown-green landscape disappeared beneath impounded water. A raft of peat broke free and rode like some stricken hulk until the waves beached it on a newly eroding shore.

The location of the Northern Water Sedge was the first to go, the first real loss in biological terms, for though material had been removed both to the herbarium and for propagation, a plant in cultivation is not the same as one growing in the wild. This loss was soon followed by that of the Tall Bog Sedge. Although the former was refound soon after, growing further down the Dale, two populations of special plants, 0.5% of the total diversity of the flowering plants of the Upper Dale, had been wiped off the face of local evolution.

Each time rain fell on the catchment the tide of death crept up and the list of losses grew, 40% of both the Alpine Rush and Rare Spring Sedge, 10% of the Rock Violet, and 5% of its rare hybrid vanished and the overall populations of many other plants were affected. We know because while it was happening Margaret Bradshaw's band of willing workers crawled and counted, marked and mapped the river's new domain.

Two hectic years rushed busily away until in January 1972 the first flow of water over the sill of the High Dam marked the fact that both the reservoir and the data banks relating to its compass were as full as they would ever be. The water and the information were there ready to be put to good use.

It would be so easy to be negative especially in hindsight of the facts. If the application to build a dam had been made in the age of conservational enlighten-ment ten years later, would the answer still have been yes? What if the water authority has waited until deep borings into aquifers beneath the industrial centres, or the water net system which could boost one river's flow by pumping from the next had been completed? The treasure of the Upper Dale could have been saved, the Great Dam no more than a white elephant of short term planning.

It would however be wrong, for evolution always looks forward and never

Brown Trout   Bullhead or Miller's Thumb

Ramshorn Snail   Great Diving Beetle

Whelk   Three-spined Stickleback

PLATE 67

back, the process of natural selection is there, ready to seize upon any new opportunity that comes its way. The message is, *think positive.*

The Man made lake is without doubt a thing of great beauty, a mirror to the fells reflecting the moods of the environment and the march of the seasons. Even in draw down it is not as bad as one might have contemplated. Silver feeder streams snake their way across the expanding apron of black peat and all the wading birds which use the moor round about for breeding and the free space above for their aerial display, do not look out of place parading in the shallows. These include no less than eleven species and if you are lucky and don't make too much noise you can see them all from the new tarmacadam road which skirts to the south bank of the lake serving the only working farm higher up the Dale on that side. Lapwing chase each other, rolling in the exuberance of another fine spring day, their chicks no more than balls of speckled fluff but each one perfectly camouflaged against the background vegetation. They are in fact most commonly seen flying above the meadows from the main road up the Dale, but some are nesting around the reservoir along with Curlew, Redshank, Dunlin, Common Sandpiper and Oystercatcher, together making full use of the diversity of environment. So much to see and so much still to preserve.

Below high water mark the old river basin lay dead, yet within that enormous weight of water there were stirrings of new life. The Brown Trout now had the freedom of still water and they began to grow much faster on a surfeit of food. This included all the small animals like Earthworms and Leather Jackets which had lived in the soils and peats before flooding, and cultures of plant plankton 'living it up' on the mineral nutrient released from the decaying terrestrial plants and animals and leached from the disintegrating soils. The Bullheads successfully extended their habitats taking up residence in the small becks feeding the upper margin of the new lake. The complexity of interrelation between a new lake and the life it will eventually support is enormous and it will be many years, perhaps hundreds, before we know how the potential of the reservoir will be exploited both by fish and sport fishermen.

Perhaps the most spectacular increase seen during and after construction of the reservoir was in the number of day visitors to the Upper Dale. The reasons were threefold. First, the increasing numbers of family car owners among the local population, mainly centred in the lowland conurbation, some of which had benefited directly from the stabilising presence of the reservoir. Second, was the publicity given to the construction of the dam by all the modern media of communication. Third, and most important of all, this period had seen a gradual awakening of interest in the natural environment and in natural history, and the development of a new concept in the social evolution of mankind, conservation. Many happenings had spurred this interest, the Silent Spring described by Rachel Carson, the *Torrey Canyon* Disaster, the *Amoco Cadiz*, mercury poisoning in Minimata Bay, Aldrin, Dieldrin, DDT, PCB, cholera in the Bay of Naples, the Sahele famines, the plight of the Indians in the forests of Amazonas, the rising price of energy and the local one, the damming of the Upper Dale, part of Britain's most important botanical heritage, to name but a spectacular few. If a single event can be pinpointed which set Man on the road to conservation it was the ultimate product of the affluence of the mid-1900s, the moment when Neil Armstrong, standing on the Moon, looked back and recognised Mother Earth for what she really is, a spaceship floating in a sea of nothingness, the life support systems of which are finite resources, there to be used not abused, for they are all we have.

The cycle of the Four Great Seasons was complete, the fullness of autumn warns of another winter soon to come. The next ice age? I think not, for there is no shred of evidence to suggest its coming and if it does there is little or nothing we can do to stall its approach. Yet there is firm evidence of another great destructive force, and that is Man himself. The pressures on this autumnal world of ours (there are already four billion of us to share its resources) are enormous and unlike the ice sheets of the last Great Winter the destructive power of Man knows no bounds. The one redeeming feature in all this is that like the glaciers of the past Man is also a constructive force and what is most important of all he alone has the ability to direct that 'construction' where he pleases.

Roe Deer Fawns with Honeysuckle, Lords and Ladies (Wild Arum)
Jacob's Ladder, Sainfoin and Elephant Hawk Moth

PLATE 68

Man is set aside above all other products of evolution by the power of conscious thought. I am myself convinced that the Whales and Dolphins share some of that same ability but I have no proof. On the evidence to hand we must conclude that we alone can make the rational decision to put aside the rat race dogma of survival of the fittest and realise that the only way ahead is to obey the true philosophy of evolution and reap the full potential of this planet Earth through co-operating, not only with our fellow human beings of every colour, creed and kind, but with all the other products of evolution.

Throughout the story of the Four Great Seasons we have never seen a single organism reign supreme within the Upper Dale. The Red Snow Algae comes nearest to the description of a successful monoculture, but even its limited success was backed by decomposer organisms who shared its frigid habitat. From that point on it was communities all the way, their make up mirroring and developing the potential then on offer. The open communities of spring were soon replaced by the forests of the Great Summer which nurtured Man the hunter gatherer, the first of the dual ecomonies which have maintained him ever since. Fossil fuel laid down by communities of the past alone allowed him to shake off his dependence, at least in part, upon the natural (forest) scene and take his first real step on the road to world domination. The fruits of that Great Autumn are now already on the wane; the stocks of many of the world's resources are running low.

If Man was any normal product of evolution without the free will to choose his course of reaction, he would have to obey the rule, 'the potential is there, it must be used'; to hell with the future. The exciting fact is that we are the first production of evolution (creation if you will) to understand our past, appreciate the present and know there is a future worth planning for.

While the energy stocks last we have much to do.

First, we must take note of *Lagopus scoticus* and contain our population in relation to the resources of the future, and note that it is a future much more distant than next year's Heather crop! Surely we have more acceptable ways of doing that than allowing the weak to go to the wall of starvation, in limbo between the resources and territories held by greater powers.

Secondly, we must move our thinking away from dependence upon fossil fuels, the deposit account of world energy, towards a greater dependence upon the current account of solar power which daily beams in from outer space. It's there to be used in the form of water, wind and wave power, as well as in the energy fixed each day by the green plants of the world.

Remember every year an average 1,045 million kilowatt hours of energy falls as rain upon the Upper Dale, the contemporary vegetation of which stores as much as 700 million kilowatt hours of energy in the same time, a figure which was more than double during the Great Summer.

Even if we look upon the future with most optimistic eyes and decide that an alternative source of unpolluting, clean, safe energy will be found, the fact remains that once we have used up all our coal, oil and natural gas, we have destroyed our major source of raw materials for the chemical industries upon which our plastic existence so depends. If we have the energy we can recycle anything we have discarded and buried as rubbish. It is much more difficult, if not impossible, to regain the products of combustion released into the atmosphere by the engines of industry, transport, or even the modern home fire. Man's future must therefore turn more and more to contemporary plant products to provide those raw materials which are being so wantonly squandered even at this moment.

We live on this earth alongside some three million other species of plants and animals which together represent the success story of 3.6 billion years of evolutionary effort. Against a background of massive global change, volcanic cataclysm, continental drift, ice ages, together they have maximised the flow of solar energy through the living landscapes of the world. They are the main heritage, from all our pasts, and our only hope for the future. We must make sure that nothing more is lost, and that goes as much for information concerning the make up and the working of the communities as for the genetic information contained within their evolving populations.

This means the setting aside of areas of the world which contain a cross section of that evolutionary effort as International Nature Reserves. They must be large enough to be viable units, for remember every community requires a certain size in which to find its natural expression. What is more, *international* funding must be found both for their institution and their upkeep, especially in the developing countries whose populations live on the brink of starvation and who cannot therefore be expected to look to long term planning.

Like the Upper Dale, the majority of the great Nature Reserves, Game Reserves, and National Parks being developed across the world are in what have until now been backwaters of human endeavour. In many cases they have only been brought to the forefront of interest in the twentieth century by the increasing lobby of conservation. Over the past twenty years conservation has become a growth industry and, here is the really bitter pill, that growth is beginning to threaten the destruction of the things it set out to conserve. Ecologists have been dragged screaming from their ivory towers of research to talk in layman's terms about the complexities of their subject which is still in its scientific infancy. Environmental consultancies have sprung up from the most unlikely sources, giving instant advice to those who can foot the bill. Specialist holidays take millionaires and back-packers alike to see it all for themselves, and even a well meaning camera tripod can become a vehicle of destruction.

Areas of the Upper Dale which have remained safe, backwaters at least in the most destructive period of Man's endeavour, have begun to come under unprecedented human pressure. The track across the head of the Dale which was closed by Hugh Baliol to stop the miners damaging the forest is now a major highway of interest shod in hiking boots. When I first knew it a mere twenty years ago it was no more than a shadow on the face of the fells; in no place more than two feet wide it meandered its way in harmony with the living landscape. Now in places, especially where it crosses the once living blanket peat, it is as much as a hundred metres across and erosion is speeding both peat and plastic litter down towards the river.

Again much of the burden and little of the benefit falls on the local people who still try to make a living from the fells. Even from the comfort of a sprung seat in a heated tractor cab, farming in the Upper Dale is still a tough job and the tramp of tourists' feet, however well orienteered, don't make it any easier. Some farmers have responded to the new potential and moved towards a new dual economy by developing caravan sites, bed and breakfast, and *en route* refreshment. Even in the best administered of these endeavours the return to the local community at large is not very great, for the visitors are able to stock up in hypermarts of the lowlands where everything is much cheaper than in the local village shops. Their main input into the local community thus becomes garbage left behind for local disposal.

The writing is fairly and squarely on the wall. Conservation (which is critical to all our futures) has got to be made part of a new economy, a real investment, not for short term gain but for the future of the world. If the affluent nations of the world don't believe in this new positive enterprise and back up that belief by the investment of meaningful sums of money, the story of The Great Seasons will turn full circle and Man will destroy not only himself but much of the living envelope that supports him. I am however an optimist and that is why this book was written, though, the pessimistic view may be that all it can achieve is an increase of interest in the Upper Dale which will hasten its destruction.

The Upper Dale is a real place and all the facts which form the basis of this story are as true as they can be within the limits of one man's knowledge at the time of writing, although I must admit a certain bias in their interpretation. For those already in the know, all they need to do is put two and two together and write a review or make another pilgrimage, I hope with deeper insight, into this very special place. For those with the will to find out it will be the work of but a few minutes to pinpoint the location of the story. They will be the members of what I like to call the converted and hence the knowledge should not be too misused. There are of course the idiot few who take birds' eggs for profit, kill anything that moves, just for the sake of killing or for mere trophy, or who dig up plants from the wild to hoard them greedily in their own private garden. I cannot understand the

The Empty Gentian Bowl

PLATE 69

mentality of these people and especially the latter, who often go on expensive tours led by an expert to see the specialist features of the flora of another country and then pillage that flora to satisfy their own powers of cheap ownership. What is more they often break the laws of their own country by illegally importing propagatable material, laws which were made to protect our native flora, garden plants and crops from disease. So often the end result is that the stolen plant dies in the soapy corner of a spongebag or forgotten in a polythene envelope. They are sick people, their sickness relating back to a belief in the rat race philosophy of the past.

The problem is that all too often natural history is regarded as something special that happens only way out in the country and its principles have nothing to do with our life at home. Pollution happens in other people's rivers, starvation is at least two worlds away and conservation only applies to the other guy.

The story of The Great Seasons is in no way remote; for you people who live in the affluence of the north temperate zone it is the story of your own backyard, the shaping of the immediate landscape which you affect by your very presence. Remember too that the same zone encompasses the major wheat belts of the world which provide the staple food for a cross section of mankind and fill the granaries of hope against the spectre of future famine; that the draw down of the oceanic water was worldwide and pollen evidence from the depths of the rain forests of the Amazon show that the effects of the Great Winter were felt clear across the tropics.

Get out and take a look, read the signs, aquaint yourself with the happenings that shaped the living landscape in which you find your roots, U shaped valleys which show the presence of the glaciers and scratches on the rocks which point out the way they went. Bend down and feel the structure of the living soil again, and look for the relicts of those plants and animal communities which formed the living landscapes of the past. Wherever you live you can look back at least to the fullness of the Great Summer before Man came upon the scene, and understand both the constructive and destructive signs of his presence.

This is the purpose of the book which is itself an amalgam of some of the knowledge and expertise of the sciences and the arts, backed by the skills of the many professions and trades which go to create, print, and bind a book, and find a market for it; to let you out from behind your air conditioned, pre-packed, double glazed, plastic coated, silicon chipped, way of life, to feel the promise of winter snow, the softness of spring rain, the warmth of summer sun, the rich blend of autumn; to let the pulse of the seasons put a sense of purpose back into your living as part of a community which spans the world. If *The Great Seasons* has done no more than this, we can ask no more.

We know a place where with care and knowledge it is possible to reach back through the whole history of a landscape and in a moment of understanding find the roots of our past, reason for our present, and sense for our future — do you?

New Life in the Upper Dale.

PLATE 70

Dramatis Plantae

Survivors from The Great Winter and Early Spring.

Group 1. Plants whose contemporary distribution centres on the Arctic.

Alpine Foxtail	*Alopecurus alpinus*
Haplodon wormskjoldi	
Lady's Mantle	*Alchemilla wichurae*

Group 2. Plants whose contemporary distribution centres on Arctic and Alpine regions.

Alchemilla filicaulis	
Alchemilla glomerulans	
Alpine Bartsia	*Bartsia alpina*
Alpine Bistort	*Polygonum viviparum*
Alpine Cinquefoil	*Potentilla crantzii*
Alpine Meadow Rue	*Thalictrum alpinum*
Alpine Timothy Grass	*Phleum alpinum*
Alpine Willow Herb	*Epilobium anagallidifolium*
Bear Berry	*Arctostaphylos uvae ursi*
Bog Sandwort	*Minuartia stricta*
Catascopium nigritum	
Cinclidium stygium	
Cloud Berry	*Rubus chamaemorus*
Dwarf Birch	*Betula nana*
False Sedge	*Kobresia simpliciuscula*
Green Spleenwort	*Asplenium viride*
Grimmia agassizi	
Hair Sedge	*Carex capillaris*

149

Hoary Whitlow Grass   *Draba incana*
Mountain Avens   *Dryas octopetala*
Northern Water Sedge   *Carex aquatilis*
Rose Root   *Sedum rosea*
Scottish Asphodel   *Tofieldia pusilla*
Spring Sandwort   *Minuartia verna*
Tall Bog Sedge   *Carex paupercula*
Three Flowered Rush   *Juncus triglumis*

Group 3. Plants whose contemporary distribution centres on the Alps.

Alpine Penny Cress   *Thlaspi alpestre*
Alpine Forget Me Not   *Myosotis alpestris*
Blue Sesleria   *Sesleria albicans*
Scapania aspera
Splachnum vasculosum
Spring Gentian   *Gentiana verna*

Group 4. Plants whose contemporary distribution centres on the mountains and the north.

Alchemilla acutiloba
Alchemilla monticola
Alchemilla subcrenata
Alpine Rush   *Juncus alpino articulatus*
Balfour's Meadow Grass   *Poa glauca*
Birds Eye Primrose   *Primula farinosa*
Camptothecium nitens
Chickweed Willow Herb   *Epilobium alsinifolium*
Globe Flower   *Trollius europaeus*
Holly Fern   *Polystichum lonchitis*
Lesser Tway Blade   *Listera cordata*
Mountain Everlasting   *Antennaria dioica*
Small White Orchid   *Leucorchis albida*

Shrubby Cinquefoil	*Potentilla fruticosa*
Tea Leaved Willow	*Salix phylicifolia*
Variegated Horsetail	*Equisetum variegatum*
Lesser Tway Blade	*Listera cordata*
Mountain Everlasting	*Antennaria dioica*
Small White Orchid	*Leucorchis albida*
Shrubby Cinquefoil	*Potentilla fruticosa*
Tea Leaved Willow	*Salix phylicifolia*
Variegated Horsetail	*Equisetum variegatum*

Group 5. Plants whose contemporary distribution centres on coastal habitats.

Sea Plantain	*Plantago maritima*
Sea Thrift	*Armeria maritima*

Survivors from the late Spring and early Summer.

Group 6. Plants whose contemporary distribution centres on the northern continent.

Bitter Milkwort	*Polygala amara*
Hairy Stonecrop	*Sedum villosum*
Heath Sedge	*Carex ericetorum*
Melancholy Thistle	*Cirsium heterophyllum*
Northern Bedstraw	*Galium boreale*
Rhytidium rugosum	
Rock Violet	*Viola rupestris*

Group 7. Plants whose contemporary distribution centres on the southern continent.

Hoary Rockrose	*Helianthemum canum*
Horseshoe Vetch	*Hippocrepis comosa*

Glossary

Scientific, Technical and unusual Words.

Albedo. Light reflected back from ice, snow, white sand etc.

Biomass. Mass (usually expressed as dry weight) of all, or a stated part, of the organic material present in a stated area.

Calcareous. Containing salts of calcium.

Carapace. Part of the external skeleton of members of the group of animals which includes insects, crabs and lobsters.

Celsius. Temperature scale °C, in which 0°C is the freezing and 100°C is the boiling point of pure water at standard pressure.

Colloidal. Refers to substances which though insoluble exist in a finely divided state thereby presenting an enormous (reactive) surface area to a solvent.

Draw down (of reservoir). When the supply of water is insufficient to keep a reservoir full it is said to be in draw down.

Drumlin. Elliptical hills found on glaciated plains, their methods of formation are complex and various.

Eskers. Ridge like landforms formed from deposits laid down by waters derived from melting ice.

Firn. Snow compacted by its own weight so that much of the air is eliminated from between the crystals.

Joule. A unit of energy, 4190 joules equals one kilocalorie.

Lotic. Refers to moving water (as a habitat).

Meristem. Living plant tissue the cells of which have the ability to divide producing more cells and hence growth.

Moraine. A landform produced by the deposition of rock debris by glacial ice.

Nunatack. Tops of mountains protruding above an ice sheet.

Palmella. A type of vegetative reproduction in which daughter cells produced by division are held in a gelatinous mass which protects them until conditions are right for further development.

Photosynthesis The physico-chemical process by which green plants convert carbon dioxide, water and sunlight into sugar and oxygen.

Phytomass Mass (usually expressed as dry weight) of all or a stated part of the plant material present in a stated area.

Refugium An area in which plants and/or animals find sanctuary in the midst of adverse conditions.

Sandar Usually large fan like expanses of rock debris (gravel, sand and silt) laid down by melt water streams issuing from glaciers.

Taiga Vegetation dominated by coniferous trees developed in a sub-arctic climate.

Taxon An abstract unit of classification, species, genus, family etc.

Taxonomy The science of classifying (grouping) things.

Till Eroded rock debris deposited by glacial ice.